Plant Breeding and Genetics
Present Concept and Approaches

THE EDITOR

Ms. Rhitisha Sood pursuing Ph.D. in Genetics and Plant Breeding (DST, INSPIRE from CSK Himachal Pradesh Krishi Vishvavidyalaya. B.Sc. (Ag.) and Masters with Kalpana Chawla Scholarship respectively. Publications 4 research papers, 18 articles, 1 popular article in Indian Farming (ICAR), 7 book chapters, 5 posters, best thesis and article award in a national conference, lifetime member -Just Agriculture.

Manoranjan Senapati is a Ph D Research Scholar and M.Sc. (Ag.) at the Division of Genetics, ICAR-Indian Agricultural Research Institute, New Delhi. He has qualified with ASRB NET inGenetics and Plant Breeding. He has been awarded ICAR- JRF-2017, CSIR -JRF-2019 and DBT- JRF-2020 in life Science. He has published articles and Book chapters in both international and national journals. He is more interested in current trends and recent development in the area of Genetics and applied research in field crops.

Ms. Shivani completed her graduation in Agriculture (Hons.) and post-graduation in discipline of Seed Science & Technology from CCS HAU, Hisar along with merit stipend of institute. Currently, she is pursuing Ph.D. in Seed Science & Technology with merit scholarship in CCS HAU, Hisar. She had cleared ASRB NET as well as contributed into 5 research paper, 2 book chapters, 7 popular articles in different agriculture magazines and 4 posters and abstracts published in international conferences. She had awarded with first prize in best poster presentation in international conference as well.

Aavula Naveen obtained B.Sc. Agriculture from Professor Jayashankar Telangana State Agricultural University, Hyderabad, Telangana and M.Sc. Genetics and Plant Breeding from Banaras Hindu University, Varanasi, Uttar Pradesh. He is currently pursuing Ph. D. Genetics and Plant Breeding from Indian Agricultural Research Institute, New Delhi. He received ICAR-JRF, ICAR-SRF fellowships during his M.Sc. and Ph.D. tenure and qualified NET(ASRB). He has contributed 2 research papers, 2 review papers, 2 books, 5 book chapters, 7 popular articles etc.

Ms. Kavita completed her graduation in Agriculture (Hons.) and post-graduation in discipline of Genetics and Plant Breeding from CCS HAU, Hisar along with merit stipend of institute. Currently, she is pursuing Ph.D. in Genetics and Plant Breeding with merit scholarship of CCS HAU, Hisar. She had cleared ASRB NET as well as contributed into 2 research paper, 2 book chapters, 12 popular articles in different agriculture magazines and 4 posters and abstracts published in international conferences.

Plant Breeding and Genetics
Present Concept and Approaches

– Editors –

Rhitisha Sood

Manoranjan Senapati

Shivani

A. Naveen

Kavita

2023

Daya Publishing House®
A Division of
Astral International Pvt. Ltd.
New Delhi – 110 002

ISBN: 9789359191638

Published by : **Daya Publishing House®**
A Division of
Astral International Pvt. Ltd.
– ISO 9001:2015 Certified Company –
47636/23, Ansari Road, Darya Ganj
New Delhi-110 002
Ph. 011-43549197, 8130496929
e-mail: info@astralint.com
Website: www.astralint.com

Printed at : **Replika Press Pvt Ltd**

Acknowledgment

Although I experienced a lot of ups and downs during this time in my life, converting one's ideas and thoughts into a book was internally rigorous yet worthwhile. Foremost I am bowed to the omnipresent "God" for providing strength, endurance and patience altogether to overcome the hurdles that crossed my way and made this writing an easy task for me. Without the experiences, efforts and assistance from my co-authors Manoranjan Senapati, Shivani Aggarwal, Aavula Naveen and Kavita this book would not have existed. I am eternally grateful to Astral Publishers for their gentle behavior and kind cooperation that helped accomplishing the publication and bringing it to the market. I am sincerely grateful to my seniors Kunal Narwal and Sunil for their generous involvement and guidance during the preparation of the manuscript of this book.

If I don't express my deep gratitude to my mother Smt. Madhu Sood, father Dr. Arun Sud, my elder brother Rhijul Sood and my beloved friend Dr. Ajay Chauhan for their steadfast devotion, selfless sacrifice, miraculous blessings, love, and support, all the words in the lexicon will be meaning less and this sweet dream could not have been sketched to reality.

Acknowledgements are inherently endless & incomplete, and I request indulgence from all the people who have given me the chance to lead them, be led by them, or observe their leadership from a distance for providing the inspiration and laid foundation for this book.

Rhitisha Sood

Preface

Genetics is the significant division of biology dealing with the study of genes, their variation and inheritance patterns among different organisms. Plant breeding, on the other hand is the genetic manipulation of plants with greater human utility which involves choosing plants that are deemed to be attractive or valuable economically or artistically, first by regulating the mating of those chosen individuals, and then by choosing particular individuals from the offspring. These procedures, when carried out repeatedly over many generations, have the capacity to alter a plant population's hereditary composition and value much beyond the range of naturally occurring populations. Thus, we can say that both the terms go hand in hand for further understandings and improvements.

The fundamental perception regarding the subject is essential and even if there is a wealth of literature and reviews available nowadays, it still takes a great deal of work to draw the appropriate conclusions. This writing of this book has been undertaken, consequently, with the conviction, by keeping the persistent priorities of students under consideration that who want to excel in the fields of plant breeding and genetics opting it as their professional carrier.

Since the conventional plant breeding methods and its systems still is the mastery in itself and cannot be replaced, but unfortunately, with rapid globalization, grooming population and its reckless demands on daily basis, consumers want the quick results in no time without considering the repercussions. To meet the requirements, plant scientists at national, international and various other firms around the world have indulged themselves especially in the molecular field labs (Marker assisted selection, gene editing/genetic engineering *etc.*) which could lead to a better understanding at genetic level, that could aid to implement it at the field level to lessen the time period and give fruitful results, the integration of both breeding systems will be an asset for the farmers to grow healthy food for masses.

Rhitisha Sood

Manoranjan Senapati

Shivani

A. Naveen

Kavita

Contents

Chapter 1

Fundamentals of Gene Concepts

Akshitha N.Y.[1]*, *Mounachari*[2], *Shwetha Yadav*[1] *and Ranjitha*[1]

[1]*Ph.D. Research Scholar, Department of Genetics and Plant Breeding, University of Agricultural Sciences, Dharwad, Karnataka*
[2]*M.Sc. (Agri), Genetics and Plant Breeding, University of Agricultural Sciences, GKVK, Bangalore, Karnataka*
**e-mail: akshithany@gmail.com*

Introduction

Based on his experiments with pea plants, an Austrian monk named Gregor Johann Mendel proposed a novel theory of inheritance in the 1860s. Prior to Mendel, most people believed that inheritance was due to a blending of parental characters where phenotype of offspring is intermediate between those of parents, similar to how mixing yellow and red paint produces a orange colour. Mendel, on the other hand, believed that heredity is the consequence of distinct units of inheritance, with each unit (or gene) acting independently in an individual's genome. The inheritance of a trait, as per this Mendelian idea, is dependent on the transmission of these units. A gene is the most fundamental physical and functional element of heredity. DNA is the building block of genes. Some genes serve as blueprints for the creation of proteins. Many genes, however, do not code for proteins. An individual receives one gene from each parent for any given trial, resulting in a pairing of two genes. Genes in humans range in size from a few hundred DNA bases to over 2 million bases. The Human Genome Project, an international research effort that aimed to discover the sequence of the human genome and determine the genes that it contained, concluded that humans have around 20,000 and 25,000 genes[1]. The majority of genes are the same in all people, although a small number of genes (less than 1 per cent of the total) differ significantly between individuals. Alleles are variants of the same gene with minor changes in their DNA base sequence. These minor distinctions add to each person's distinct physical characteristics.

Classical Genetics

Our forefathers used genetics long before scientists gained interest in heredity. Human civilization was greatly facilitated by the use of genetics to domesticate plants and animals. The modern domesticated maize (*Zea mays*) is the consequence of years of artificial selection through controlled mating of ancestral Teosinte species[2]. Same is true in case of animals. Years of selective plant and animal breeding have shown that desirable features might be selected through controlled mating. For millennia, humans have puzzled how characters are passed down from generation to generation and why some traits skip generations and appear later. It wasn't until 1865, however, that people began to comprehend the process underlying this phenomena, which is largely attributable to the meticulous and exhaustive study of an Austrian monk named Gregor Mendel.

Gregor Johann Mendel (1822-1884), known as the Father of Genetics, was an Austrian monk. In 1865 he published the results of hybridization experiments titled Experiments on Plant Hybridization in a journal 'The proceeding of the Brunn society of natural history' and postulated the principles of inheritance which are popularly known as Mendel's laws[3]. But his work was largely ignored by scientists at that time. In 1900 the work was independently rediscovered by three biologists - Hugo de Vries of Holland, Carl Correns of Germany and Erich Tschermak of Austria. Mendel did a statistical study since he had a mathematical background. He discovered that individual traits are inherited as discrete factors which retain their physical identity in a hybrid. Later, these factors came to be known as genes. The term was coined by Danish botanist Wilhelm Johannsen in 1909. A gene is defined as a unit of heredity that may influence the outcome of an organism's traits.

Gregor Johann Mendel and his Experiments

Mendel was born in 1822 near Brunn in Austria, in the family of a poor farmer. Due to the financial problems, he was unable to continue his studies and hence joined the St. Augustinian Monastery at Brunn in 1843. In 1847, he was made priest. In 1851, he was sent by the monastery authorities to the University of Vienna where he studied physics, mathematics and philosophy *etc.* After completing his studies, he returned to Brunn in 1854 where he was appointed as a substitute science teacher. In addition, he worked as a priest in the local church and lived in a house located within the premises of the church. He began to collect pea seeds for his experiments in 1857 from commercial seed growers all over Europe. He conducted all his experiments within the kitchen garden of his house with the help of his own resources. He carried out seven years of painstaking, sincere, devoted experimentation. He presented his findings before the National History Society of Brunn at two of its meetings on Feb. 8 and March 8, 1865. This paper entitled, 'Experiments in Plant Hybridization', was published in the annual proceedings of the society in 1866, which was distributed to many libraries of Europe and North **America**[2]. This work of mendel remained unnoticed and did not get much appreciation. Later, Mendel became more and more involved with the work of monastery and continued his studies on honey bees, some other plants and on climatology. He died in 1884 at an age of 62 years. In 1900, sixteen years after Mendel's death, three scientists, Hugo de Vries in Holland, Correns in Germany and Tschermak in Austria, arrived at the same conclusions as those of Mendel. These three scientists, worked in different countries rediscovered the paper by Mendel. After this discovery, there was a spurt of interest in the Mendel's findings. The science of genetics thus truly born.

Mendel's discoveries enabled subsequent scientists to anticipate the expression of characteristics using mathematical probabilities. Mendel's decision to begin his crossings solely with plants that he established were true-breeding contributed significantly to his success. He also measured only absolute (binary) properties of the offspring, such as colour, shape, and location, rather than quantitative characteristics. He numerically expressed his findings and statistically analysed them. His data was credible because of his data processing method and huge sample size. He also had the insight to observe and document the changes in numerous succeeding generations (F_2, F_3) of his pea plants. Finally, he conducted "test crosses" (backcrossing offspring of the first hybridization to the original true-breeding lines) to determine the existence and proportion of recessive traits. Mendel's work would not have had the same impact on genetics without his meticulous attention to method and detail.

Selection of Pea as an Experimental Material

The choice of pea for hybridization by Mendel offered the some advantages as an experimental material. In the available varieties, several characters had two contrasting forms which were easily distinguishable from each other. This permitted easy classification of F_2 and F_3 progeny from various crosses into clear cut crosses. The flower structure of pea ensures self pollination, which was experimentally verified by Mendel. Since the flowers are relatively large, emasculation and pollination of pea flowers was easy. The duration of pea crop is of single season which helped him to conduct experiments for many generations to get accurate results. Pea seeds are large and had no problem in germination, plants are relatively easy to grow and each plant occupies only a small space. In addition to pea, Mendel also worked on rajma (*Phaseolus vulgaris* L.) as well. The results from the experiments on rajma were reported along with those on pea in the **same paper**.

Table 1.1: Characters Studied by Mendel in Pea Plant

Characters	*Dominant Form*	*Recessive Form*
Seed shape	Round	Wrinkled
Seed coat colour	Grey	Yellow
Cotyledon color	Yellow	Green
Pod color	Green	Yellow
Pod shape	Full	Constricted
Position of flower	Axial	Terminal
Length of stem	Tall	Dwarf

Allele

A gene is a section of DNA that defines a specific trait, and its exact location on a chromosome is referred to as a locus. Each gene can exist in many forms known as alleles, which code for distinct forms of an inherited trait. Alleles can alternatively be defined as genes that occupy equivalent locations on homologous chromosomes and control the same trait (*e.g.* plant height) but produce distinct effects (tall or short). The term homologous refers to chromosomes that have the same set of genes in the same order, however they may not have identical alleles of each gene.

Wild-Type and Mutant Alleles

Prevalent alleles in a population are called wild-type alleles. These alleles typically encode proteins that are made in the right amount and function normally. Alleles that are present at less than 1 per cent in the population and have been altered by mutation are called mutant alleles. Such alleles usually result in a reduction in the amount or function of the wild-type protein and are most often inherited in a recessive fashion.

Dominant and Recessive Alleles

A dominant allele is an allele that masks or hides the expression of a recessive allele. it is represented by an uppercase letter. A recessive allele is one that has an effect only when homozygous, and its expression is concealed by a dominant allele when heterozygous. A lowercase letter is used to represent it.

Homozygous and Heterozygous Condition

Homozygous and heterozygous are terms that are used to describe allele pairs. Individuals carrying two identical alleles (RR or rr) are known as homozygous. While individual organisms bearing different alleles (Rr) are known as heterozygous.

Homozygous dominant genotypes possess two dominant alleles for a trait (TT).

Homozygous recessive genotypes possess two recessive alleles for a trait (tt).

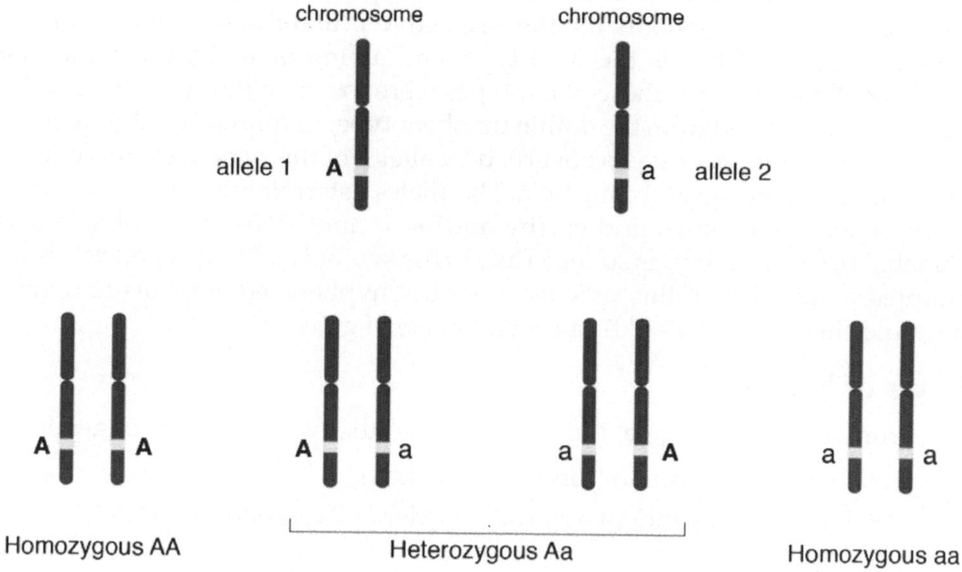

Genotype and Phenotype

Two terms are used to distinguish physical appearance from genetic constitution *i.e.* genotype and phenotype. The genotype is defined as an individual's genetic constitution for any given trait. An individual's genotype is commonly represented by a symbol, such as tt, Tt, or TT. The phenotype is an individual's observable properties for any given trait. An individual's phenotype is largely determined by his or her genetic constitution.

Ex: Height (tall/dwarf) is the phenotype where as the genes controlling the height (TT/Tt/tt) represents its genotype.

Derivation of Gene Symbols

It is customary to derive gene symbols from the names of the characters they govern. This practice is due to Bateson who, for example, used T and t to denote the alleles responsible for tall and dwarf plants, respectively, in pea[4]. Gene symbols are derived from the following: (1) name of the dominant phenotype (this was followed initially), (2) name of the recessive phenotype (it was followed subquently), (3) name of the mutant phenotype (it is followed currently), and (4) features of the encoded protein. In fact, today there are many specialized systems for deriving gene symbols, *e.g.*, for genes of bacteria and viruses. These different systems have evolved in response to the demands posed by new discoveries.

In case of multiple alleles, symbols for the wild type and the first discovered mutant allele are derived as described earlier. Symbols for mutant alleles discovered subsequently are obtained by adding the first letter or the first letter and one or two more letters from the respective mutant phenotypes as superscripts to the symbols for the respective mutant alleles. For example, coloured fur in rabbits is the wild type, and albino or white fur is the first discovered mutant type; these phenotypes were assigned the symbols C and c, respectively (derived from the dominant phenotype, 'coloured"). Subsequently, two more mutant phenotypes governed by alleles of this gene were discovered and named himalayan and chinchilla. The alleles determining these phenotypes were given the symbols and ch (by adding h and ch from himalayan and chinchilla), respectively, as superscripts to the symbol c). Plant geneticists have adopted a variation of this practice; they use hyphenated symbols to denote multiple alleles, *e.g.*, sh2-6801 (a mutant allele of gene sh2).

Laws of Mendel

From his studies in pea, Mendel proposed the two basic laws of genetics:

a) Law of segregation (Mendel's first law)

b) Law of independent assortment (Mendel's second law)

1. Law of Segregation

The law of segregation states that when a pair of contrasting alleles are brought together in a heterozygote (hybrid) the two members of the allelic pair remain together without being contaminated and when gametes are formed from the hybrid, the two alleles separate out from each other and only one enters each gamete.

Example: Pure tall plants are homozygous and so have genes TT. Similarly, dwarfs have tt genes. Tallness and dwarfness are two distinct yet opposing factors. Tall plants produce gametes with gene T, while dwarf plants produce gametes with gene t. During cross fertilisation, gametes containing T and t combine to form F1 hybrids. These hybrids have the genotype Tt. It indicates that, despite being tall phenotypically, F1 plants have one allele for tallness and one gene for dwarfness. The tall and dwarf characters appear to have become contaminated, resulting in only the tall character. However, during gamete production, the genes T (for tallness) and t (for dwarfness) split and are passed on to different gametes. As a result, the heterozygote produces an equal quantity of two types of gametes. 50 per cent of the gametes have allele T, whereas the other 50 per cent have allele t. As a result, these gametes remain pure for both tallness and dwarfness. This is why the law of segregation is also known as the law of gamete purity. Gametes combine at random, and when gametes are numerous, all possible combinations can occur, resulting in a 3:1 ratio of tall to dwarf. Punnett square is used to interpret these results.

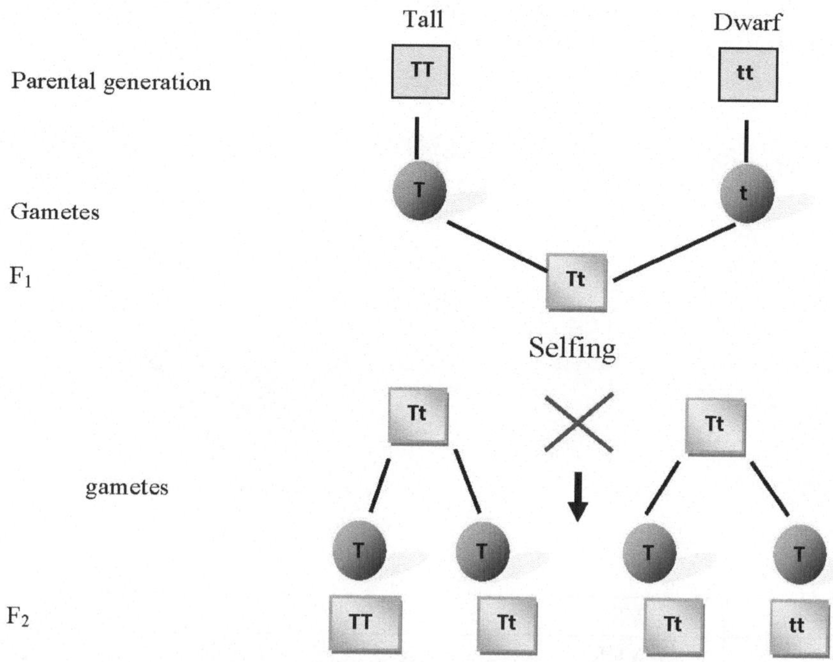

Genotypic ratio: 1: 2: 1, Phenotypic ratio: 3:1

2. Law of Independent Assortment

This law describes how different genes independently separate from one another when reproductive cells develop. Mendel developed this law based on the findings of a dihybrid cross. The law of independent assortment asserts that alleles of various genes are inherited independently within sexually reproducing organisms.

Explanation: In a dihybrid cross, the cross was made between plants having yellow and round cotyledons and plants having green and wrinkled Cotyledons. The F_1 hybrids all had yellow and round seeds. When these F_1, plants were self fertilized they produced four types of plants in the following proportion.

(i) Yellow and round-9

(ii) Yellow and wrinkled-3

(iii) Green and round - 3

(iv) Green and wrinkled-1

Above results indicate that yellow and green seeds appear in the following ratios: 9+3: 3+1 = 12:4 or 3:1. Similarly, the ratio of round and wrinkled seeds is 9+3:3+1 = 12:4 or 3:1. This suggests that each of the two pairs of alternative characters, namely yellow-green cotyledon colour, is inherited independently of the cotyledons' round-wrinkled character. It suggests that the factor for yellow colour enters the gametes independently of R or r, implying that gene Y can be passed on to the gametes either R or r.

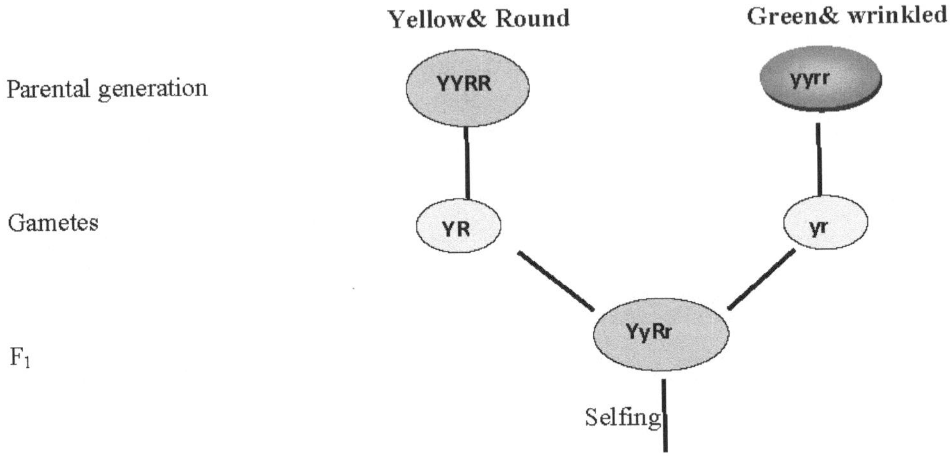

gametes	YR	Yr	yR	yr
YR	YYRR	YYRr	YyRR	YyRr
Yr	YYRr	YYrr	YyRr	Yyrr
yR	YyRR	YyRr	YyRR	yyRr
yr	YyRr	Yyrr	Yyrr	yyrr

Genotypic ratio: 1: 2: 2: 4: 1: 2: 1: 2: 1

Phenotypic ratio: 9: 3: 3:1

Test Cross

A test cross is a hybrid between an F1 individual and a homozygous recessive parent, while a back cross is a cross between an F1 individual with one of the parental genotypes. Test cross is performed to determine the genotype of an individual with dominant phenotype (*i.e.* homozygous or heterozygous).

Monohybrid Test Cross

Here the F1 of monohybrid cross is crossed with recessive parent. Consider the case in which a test cross is performed with a dwarf pea plant (recessive) and a tall pea plant of unknown genotype

(a) The tall pea plants are homozygous

Tall (TT) x Dwarf (tt)

Gamtets Only T ↓ Only t

F1 Tt (all tall)

Conclusion: since F1 contain only tall plants, dominant parent must producing only one type of gamete and thus, it is homozygous dominant(TT)

(b) The tall pea plants are herozygous

	Tall (Tt)	x	Dwarf (tt)
Gamtets	T and t	↓	Only t
F1	1Tt (tall): 1 tt (dwarf)		

Conclusion: since F1 contain both tall and dwarf plants, dominant parent must producing two types of gametes and thus, it is heterozygous dominant(Tt)

Dihybrid Test Cross

A dihybrid test cross with a double heterozygote (YyRr) yields a genotypic and phenotypic ratio of 1: 1: 1: 1, suggesting that two sets of components segregate and assort independently.

Parents	Yellow and Round	x	Wrinkled and Green
Genotype	(YyRr)	|	(yyrr)
Gametes	YR, Yr, yR, yr	↓	only yr
F1		YyRr (1/4)	
		yyRr (1/4)	
		Yyrr (1/4)	
		yyrr (1/4)	

Phenotypic ratio and genotypic ration: 1:1:1:1

Rediscovery of Mendel's Work

Mendel's findings were largely ignored. The importance placed by nineteenth-century biologists on the apparent blending of inherited traits in the overall appearance of the progeny, as opposed to the organ-specific binary characters studied by Mendel, was a major impediment to understanding their significance. In 1900, however, his work was "re-discovered" by three European scientists, Hugo de Vries, Carl Correns, and Erich von Tschermak[5]. The precise nature of the "re-discovery" has been debated: De Vries was the first to publish on the subject, mentioning Mendel in a footnote, while Correns pointed out Mendel's priority after reading De Vries' paper and realising that he himself did not have priority. De Vries may not have acknowledged honestly how much of his knowledge of the rules came from his own work or only after reading Mendel's publication. Later researchers accused Von Tschermak of not properly comprehending the results. Regardless, the "rediscovery" elevated Mendelism to the status of an important but controversial theory. Its most ardent promoter in Europe was William Bateson, who coined the terms "genetics," "gene," and "allele" to characterise several of its ideas.

Complete Dominance

In case of complete dominance, the phenotype of heterozygotes is identical with that of homozygotes for the concerned dominant allele. The dominant allele in such a situation is known as completely or fully dominant. An example of complete dominance occurs for seed shape in peas. Characters showing complete dominance yield the typical 3: 1 monohybrid ratio in F_2.

Incomplete Dominance

Incomplete dominance (also known as partial dominance) arises when the phenotype of the heterozygous genotype differs from, and frequently intermediates between, the phenotypes of the homozygous genotypes. The snapdragon flower colour, for example, is either homozygous for red or white. When the red homozygous flower is crossed with the white homozygous flower, a pink snapdragon flower is produced. Pink snapdragons are the result of incomplete dominance. When F_1 generation plants are self-pollinated, the phenotypic and genotypic ratio of the F_2 generation is 1:2:1. (Red: Pink: White).

Codominance

Contributions of both the alleles can be seen in case of codominance. In heterozygotes, both alleles of a gene express themselves. As a result, heterozygotes for such genes exhibit the traits caused by both alleles. A and B blood types in humans presents an excellent example of codominance. Chemical modifications to a glycoprotein (the H antigen) on the surfaces of blood cells are controlled by three alleles in the ABO blood group system, two of which are co-dominant (I^A, I^B) and dominant over the recessive i at the ABO locus. Thus, I^A and I^B alleles are each dominant to I (I^A I^A and I^Ai people have type A blood, whereas I^B I^B and I^Bi people have type B blood). However, I^A I^B individuals have both alterations on their blood cells and hence have type AB blood, indicating that the I^A and I^B alleles are co-dominant.

Lethal Genes

Gene, which causes the death of its carrier when in homozygous condition is called lethal gene. Mendel's findings were based on equal survival of all genotypes. In normal segregation ratio of 3:1 is modified into 2:1 ratio.

Types of Lethal Genes

Dominant Lethal

Lethality of the individual in both homozygous (Fully dominant lethal) and heterozygous condition. The individuals with dominant lethal die before they can leave progeny. Therefore the mutant dominant lethal is removed from the population and the expected phenotypic ratio is altered.

In mice, yellow colour is produced by dominant gene (Y), while its recessive allele y determines the normal gray coat colour. Further, all the mice with Yellow coat colour were heterozygotes(Yy) and was unable to find a mouse homozygous for the Y allele (YY) and that it causes the death of homozygous YY embryos at an early stage of development.

Recessive Lethal

Show lethal effect only when they are in homozygous recessive state and they exhibit distinct phenotypes.

Examples: Albinism in plants, Sickle cell anaemia.

Conditional Lethals

Some lethal genes allow normal development and survival, except under specific condition which is necessary for their lethal effect. The disease Xeroderma pigmentosum, skin cancer, is produced when a person is exposed to sunlight.

Gametic Lethals

Some genes make the gametes incapable of fertilization. This lead to drastic departure from typical ratio in segregating generation. This phenomenon is commonly known as segregation distortion or meiotic drive.

Balanced Lethals

The balancing effect between two different lethals is called balanced lethal system. Lethal genes linked in repulsion phase of linkage are said as balanced lethals. They are maintained in repulsion phase due to tight linkage. Crossing over is very low. In coupling phase, it causes death.

Types of Lethal Genes - Based on Survivality

Vital genes: Those genes which do not affect the survival of the individuals in which they are present. Wild type allele of all the genes are considered as vital genes. They serve as reference point in determining the effect of mutant alleles.

Super vital genes: Some mutant allele enhance the survival of those individuals that carry them in the appropriate genotype as compared to that of the wild type allele. Super-vital genes protect the individuals carrying them against the various disease thus increasing the chance of their survival.

Semilethal or sublethal genes: Such genes do not lead the organism to the death that carry them in appropriate genotype. 90 per cent of the individuals die, however, only less than 10 per cent of the individuals survive. Eg.: Xantha mutants of several plants are sub lethal in the homozygous state. **Sub-vital**

genes: Most of the mutant genes reduce the viability of individuals having them in appropriate genotype as compared to that of normal individuals. Most of the mutant genes are sub-vital in their effect and kill less than 90 per cent of the individuals which carry them. e.g.: Virdis mutants of barley, miniature wings in Drosophila.

Lethal genes: Causes the death of an individual when carried in an appropriate genotype

Concept of Polygenes

Colour, sex *etc.* which shows distinct categories are known as qualitative characters. They are usually governed by one or major genes or oligogenes. Characters like length of ear in corn, yield of grain, yield of milk, stature *etc.* do not fall into clear cut classes and shown more or less continuous variation and are governed by a large number of minor genes called multiple genes or polygenes. Polygens show continuous variation and have a marked influence of the environment on their expression.

Multiple Alleles

In each of the seven pairs of characters studied by Mendel, there were only two alternative forms for each character. When more than two different forms of a given gene exist in a species, they are referred to as multiple alleles. A classical example of multiple alleles is found in the ABO blood group system of human which is controlled by three alleles I^A, I^B and i. However alleles are coded by single gene. The RBCs of people with type A blood group contain A antigens, those with type B blood group contain B antigens, those with type AB blood group have both A and B antigens and those with type O blood groups do not have any A or B antigens. The I^A and I^B alleles are responsible for the production of A and B antigens respectively. The allele I^a for the A antigen is codominant with the allele I^B for the B-antigen. Both I^A and I^B are completely dominant to the allele i, which fails to specify A or B antigens. The hierarchy of dominance relationship is symbolized as $(I^A\text{-}I^B) > i$.

Pleiotropy

Pleiotropy comes from the Greek terms pleio, which means "many" and tropic, which means "affecting". Pleiotropic genes are those that affect many, apparently unrelated phenotypes[6]. Pleiotropy refers to the phenomenon of a single gene determining numerous unique features. Pleiotropy should not be confused with polygenic characteristics, which occur when numerous genes interact to produce a single phenotype. Pleiotropy refers to a single gene's hereditary effect on numerous phenotypic features.

Example: Sickle cell disease

It is a genetically transmitted haemolytic disease. Homozygotes have two copies of abnormal gene (beta haemoglobin) leading to formation crescent shaped or sickle shaped hemoglobin (Hbs). These cells have very low oxygen carrying capacity. Thus impair circulation. This often led to renal failure, cardiac failure and thrombosis. Thus a single mutation in a gene has multiple consequences in numerous tissues. But in heterozygotes (having one abnormal copy and one normal copy), there is an advantage called heterozygote advantage. These heterozygotes are protected against most lethal form of malaria.

Chromosomal Theory of Inheritance

Boveri and Sutton presented the chromosomal theory of inheritance in the early 1900s. It is the basic theory of genetics. According to this theory g enes, which are the units of inheritance, are present in chromosomes. Mendelian genetics gave way to the Chromosomal Theory of Inheritance. This hypothesis describes how cellular chromosome transmission transfers genetic deternaminants (genes) from parent to offspring.The Chromosomal Theory of Inheritance supports Mendel's laws. Listed below are the observations of this theory:

(i) During the process of cell division-meiosis, the pairs of homologous chromosome move as discrete structures, which are independent of other pairs of chromosomes.

(ii) There is a random distribution of chromosomes into the pre-gametes from each homologous pair.

(iii) Each parent synthesizes gametes, which constitute only half of their chromosomal complement.

(iv) Even though female (egg) and male (sperm) gametes differ in their size and morphology, they have the same number of chromosomes, submitting equal genetic contributions from each parent.

(v) The gametic chromosomes fuse during fertilization to produce offspring with the same number of a chromosome as their parents.

Penetrance and Expressivity

The strict correlation between genotype and phenotype does not always hold true. For example, say everyone in a population of 100 individuals carries the same genotype for a certain trait (suppose polydactyly), yet only 85 individuals of the population actually shows the phenotype expected from that genotype. The proportion of genotypes that actually show expected phenotypes is called penetrance. In this example, the penetrance is 85 per cent. In general, when we know that the genotype is present but the phenotype is not observable, the trait shows incomplete penetrance. In incomplete penetrance, a genotype is not expressed by every organism in which it is present. Basically, anything that shows less than 100 per cent penetrance is an example of incomplete penetrance.

The degree of expression of a trait is controlled by a gene. A particular gene may produce different degrees of expression in different individuals. This

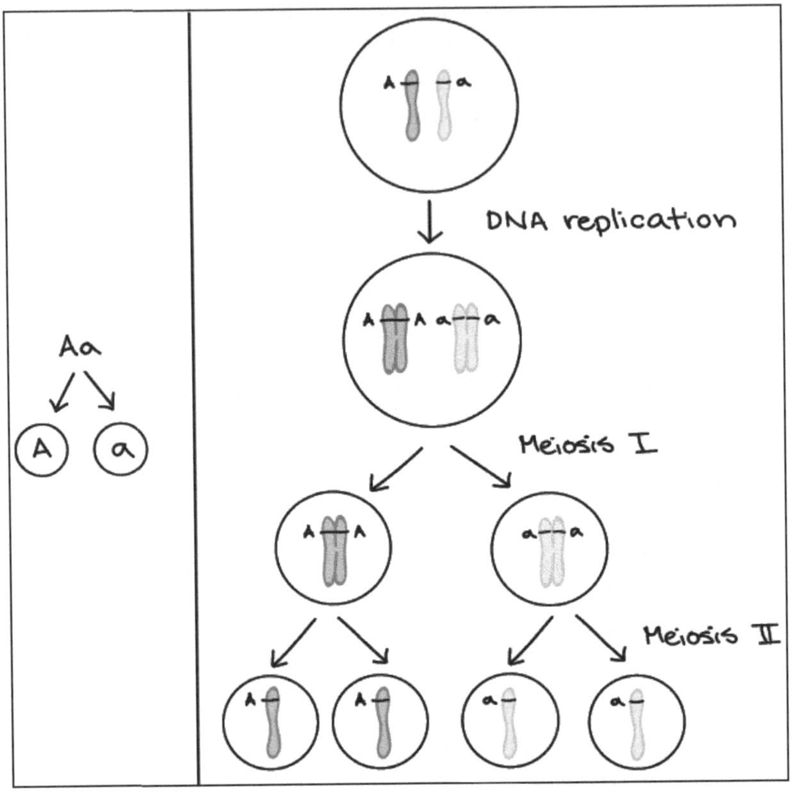

Segregation of homologous chromosome during meiosis explains Mendel's law of segregation. During the first meiotic division, homologous chromosomes pair. One of the homologs comes from the mother, the other from the father. In the anaphase of the first meiotic division, the paired chromosomes segregate and move to opposite poles of the cell. One carries allele A and the other allele a. This physical separation of the two chromosomes segregates the alleles from each other; eventually, they will reside in different daughter cells. Mendel's principle of segregation is therefore based on the separation of homologous chromosomes during the anaphase of the first meiotic division. Random alignment of bivalents during metaphase of meiosis I explains the mendals law of independent assortment.

is known as expressivity. Expressivity is the degree to which trait expression differs among individuals[7]. For example, the features of Marfan syndrome vary widely; some people have only mild symptoms, whereas others also experience life-threatening. The relationship between genotype and phenotype is not simple. The terms penetrance and expressivity quantify the modification of gene expression by varying environment and genetic background; they measure respectively the percentage of cases in which the gene is expressed and the level of expression.

Changing Concept of Gene

The concept of gene has been the focal point of study from the beginning of twentieth century to establish the basis of heredity. The gene has been examined from two main angles, *i.e.* (1) genetic view, and (2) biochemical and molecular view.

These aspects are briefly described below:

1. Genetic View

The genetic view or perspective of gene is based mainly on the Mendelian inheritance, chromosomal theory of inheritance and linkage studies. Mendel used the term factors for genes and reported that factors were responsible for transmission of characters from parents to their offspring. Sutton and Boveri (1903) based on the study of mitosis and meiosis in higher plants established parallel behaviour of chromosomes and genes. They reported that both chromosomes and genes segregate and exhibit random assortment, which clearly demonstrated that genes are located on chromosomes.

Morgan based on linkage studies in Drosophila reported that genes are located on the chromosome in a linear fashion[8]. Some genes do not assort independently because of linkage between them. He suggested that recombinants are the result of crossing over. The crossing over increases if the distance between two genes is more. The number of linkage group is the same as the number of chromosomes. The chromosome theory and linkage studies reveal that genes are located on the chromosomes. This view is sometimes called as bead theory.

The important points about the bead theory are given below:

1. The gene is viewed as a fundamental unit of structure, indivisible by crossing over. Crossing over occurs between genes but not within a gene.
2. The gene is considered as a basic unit of change or mutation. It changes from one allelic form to another, but there are no smaller components within a gene that can change.
3. The gene is viewed as a basic unit of function. Parts of a gene, if they exist, cannot function.

The chromosome has been viewed merely as a vector or transporter of genes and exists simply to permit their orderly segregation and to shuffle them in recombination. The bead theory is no more valid for any of the above three points.

Now evidences are available which indicate that:

(i) A gene is divisible,

(ii) Part of a gene can function.

i. The Gene is Divisible

Earlier it was believed that gene is a basic unit of structure which is indivisible by crossing over. In other words, crossing over occurs between genes but not within a gene. Now, intragenic recombination has been observed in many organisms which indicates that a gene is divisible.

The intragenic recombination has following two main features:

1. It occurs with rare frequency so that a very large test cross progeny is required for its detection. Benzer expected to detect a recombination frequency as low as 10^{-6}, the lowest he actually found was 10^{-4}(0.01 x 2 = 0.02 per cent).
2. The alleles in which intragenic recombination occurs are separated by small distances within a gene and are functionally related.

Examples of intragenic recombination include bar eye, star asteroid eye and lozenge eye in Drosophila.

Bar Eye in Drosophila

The first case of intragenic recombination was recorded in Drosophila for bar locus which controls size of eye. The bar locus contains more than one unit of function. The dominant bar gene in Drosophila produces slit like eye instead of normal oval eye. Bar phenotype is caused by tandem duplication of 16A region in X chromosome, which results due to unequal crossing over. The flies with different dose of 16A region have different types of eyes. Single copy of 16A region produces wild type (oval eye) where as double and triple 16A regions produces bar eye (small in size) and ultra bar eye (very small in size) respectively.

ii. Part of a Gene Can Function

It was considered earlier that gene is the basic unit of function and parts of gene, if exist, cannot function. But this concept has been outdated now. Based on studies on rll locus of T4 phage, Benzer (1955) concluded that there are three sub divisions of a gene, *viz.*, recon, muton and cistron.

These are briefly described below:

a. **Recon:** Recons are the regions (units) within a gene between which recombination's can occur, but the recombination cannot occur within a recon. There is a minimum recombination distance within a gene which separates recons. The map of a gene is completely linear sequence of recons.

b. **Muton:** It is the smallest element within a gene, which can give rise to a mutant phenotype or mutation. This indicates that part of a gene can mutate or change. This disproved the bead theory according to which the entire gene was to mutate or change.

c. **Cistron:** It is the largest element within a gene which is the unit of function. This also knocked down the bead theory according to which entire gene was the unit of function. The name cistron has been derived from the test which is performed to know whether two mutants are within the same cistron on in different cistrons. It is called cis-trans test which is described below.

2. Biochemical View

It is now generally believed that a gene is a sequence of nucleotides in DNA which controls a single polypeptide chain. The different mutations of a gene may be due to change in single nucleotide at more than one location in the gene. Crossing over can take place between the altered nucleotides within a gene.

Since the mutant nucleotides are placed so close together, crossing over is expected within very low frequency. When several different genes which affect the same trait are present so close that crossing over is rare between them, the term complex locus is applied to them. Within the nucleotide sequence of DNA, which represents a gene, multiple alleles are due to mutations at different points within the gene.

Fine Structure of Gene

Benzer, in 1955, divided the gene into recon, muton and cistron which are the units of recombination, mutation and function within a gene. Several units of this type exist in a gene. In other words, each gene consists of several units of function, mutation and recombination. The fine structure of gene deals with mapping of individual gene locus. This is parallel to the mapping of chromosomes. In chromosome mapping, various genes are assigned on a chromosome, whereas in case of a gene several alleles are assigned to the same locus. The individual gene maps are prepared with the help of intragenic recombination.

Since the frequency of intragenic recombination is extremely low, very large population has to be grown to obtain such rare combination. Prokaryotes are suitable material for growing large population[10]. In Drosophila, 14 alleles of lozenge gene map at four mutational sites which belong to the same locus (Green, 1961). Similarly, for rosy eye in Drosophila, different alleles map at 10 mutational sites of the same locus.

Descriptions about each Genes

There are some genes which are different from normal genes either in terms of their nucleotide sequences or functions. Some examples of such genes are split gene, jumping gene, overlapping gene and pseudo gene.

1. Split Genes

Usually a gene has a continuous sequence of nucleotides. In other words, there is no interruption in the nucleotide sequence of a gene. Such nucleotide sequence codes for a particular single polypeptide chain. However, it was observed that the sequence of nucleotides was not continuous in case of some genes; the sequences of nucleotides were interrupted by intervening sequences.

Such genes with interrupted sequence of nucleotides are referred to as split genes or interrupted genes. Thus, split genes have two types of sequences, *viz.*, normal sequences and interrupted sequences.

i. Normal Sequence: This represents the sequence of nucleotides which are included in the mRNA which is translated from DNA of split gene. These sequences code for a particular polypeptide chain and are known as exons.

ii. Interrupted Sequence: The intervening or interrupted sequences of split gene are known as introns. These sequences do not code for any peptide chain. Moreover, interrupted sequences are not included into mRNA which is transcribed from DNA of split genes. The interrupted sequences are removed from the mRNA during processing of the same. In other words, the intervening sequences are discarded in mRNA as they are non-coding sequences. The coding sequences or exons are joined by ligage enzyme.

2. *Jumping Genes*

Generally, a gene occupies a specific position on the chromosome called locus. However, in some cases a gene keeps on changing its position within the chromosome and also between the chromosomes of the same genome. Such genes are known as jumping genes or transposons or transposable elements.

The first case of jumping gene was reported by Barbara McClintock in maize as early as in 1950. However, her work did not get recognition for a long time like that of Mendel[3]. Because she was much ahead of time and this was an unusual finding, people did not appreciate it for a long time. This concept was recognized in early seventies and McClintock was awarded Nobel prize for this work in 1983. Later on transposable elements were reported in the chromosome of E. coli and other prokaryotes. In E. coli, some DNA segments were found moving from one location to other location. Such DNA segments are detected by their presence at such a position in the nucleotide sequence, where they were not present earlier.

3. *Overlapping Genes*

Earlier it was believed that a nucleotide sequence codes only for one protein. Recent investigations with prokaryotes especially viruses have proved beyond doubt that some nucleotide sequences (genes) can code for two or even more proteins.

The genes which code for more than one protein are known as overlapping genes. In case of overlapping genes, the complete nucleotide sequence codes for one protein and a part of such nucleotide sequence can code for another protein.

Overlapping genes are found in tumor producing viruses such as x X 174, SV 40 and G4. In virus xX 174 gene A overlaps gene B. In virus SV 40, the same nucleotide sequence codes for the protein VP 3 and also for the carboxyl-terminal end of the protein VP2. In virus G4, the gene A overlaps gene B and gene E overlaps gene D. The gene of this virus also contains some portions of nucleotide sequences which are common for gene A and gene C.

4. Pseudogenes

There are some DNA sequences, especially in eukaryotes, which are non-functional or defective copies of normal genes. These sequences do not have any function. Such DNA sequences or genes are known as pseudogenes. Pseudogenes have been reported in humans, mouse and Drosophila. The well-known examples of pseudogenes are alpha and beta globin pseudogenes of mouse.

Conclusion

Gene is structural and functional unit of chromosome. It was earlier known as factor. The term "gene" was given by Johannsen. Alternative forms of gene is known as allele and fine structure of gene revealed that, it have three subdivisions *viz.*, Cistron, muton and recon. From his studies in pea, Mendel proposed the two basic laws of genetics, Law of segregation, and Law of independent assortment based on gene concept. The chromosomal theory of inheritance was given by Boveri and Sutton, genes are the units of heredity and are found in the chromosomes. Morgan based on linkage studies in Drosophila reported that genes are located on the chromosome in a linear fashion. There are different functions of gene in the biological entity.

REFERENCES

1. Snustad D.M., and Simmons M.J., Principles of genetics. Edn 6, Vol. I, John Wiley and Sins, Inc. 2012, 1-3.
2. Yamasaki, M., Wright, S.I. and McMullen, M.D., 2007. Genomic screening for artificial selection during domestication and improvement in maize. Annals of Botany, 100(5), pp.967-973.
3. Dunn, L.C., 1965. Mendel, his work and his place in history. Proceedings of the American Philosophical Society, 109(4), pp.189-198.
4. Priyadarshan, P.M., 2019. PLANT BREEDING: classical to modern.
5. Van Dijk, P.J., Weissing, F.J. and Ellis, T.N., 2018. How Mendel's interest in inheritance grew out of plant improvement. Genetics, 210(2), pp.347-355.

6. Waters, C.K., 1994. Genes made molecular. Philosophy of Science, 61(2), pp.163-185.
7. Taeubner, J., Wieczorek, D., Yasin, L., Brozou, T., Borkhardt, A. and Kuhlen, M., 2018. Penetrance and expressivity in inherited cancer predisposing syndromes. Trends in cancer, 4(11), pp.718-728.
8. Weiling, F., 1991. Historical study: Johann Gregor Mendel 1822–1884. American journal of medical genetics, 40(1), pp.1-25
9. Ravindran, S., 2012. Barbara McClintock and the discovery of jumping genes. Proceedings of the National Academy of Sciences, 109(50), pp.20198-20199.
10. Treangen, T.J., Abraham, A.L., Touchon, M. and Rocha, E.P., 2009. Genesis, effects and fates of repeats in prokaryotic genomes. FEMS microbiology reviews, 33(3), pp.539-571.

Chapter 2

Cytogenetics and its Relevance in the Modern Breeding Era

Tushar Arun Mohanty[1]*, Tanisha Nayak[2], M. Himakara Datta[1] and Poojashree Mahapatra[3]

[1]Ph.D. Scholar (Genetics and Plant Breeding),
[2]M.Sc scholar (Genetics and Plant Breeding),
Centre for Plant Breeding and Genetics, Tamil Nadu Agricultural University, Coimbatore – 641003, Tamil Nadu
[3]M.Sc. Scholar (Plant Pathology), College of Agriculture, Odisha University of Agriculture and Technology, Bhubaneswar – 751003, Odisha
**e-mail: tushararunmohanty@gmail.com*

Introduction

The advent of plant cytology, which eventually gave rise to plant cytogenetics, was one of the most significant developments in the history of genetics and plant breeding. Research and diagnostics in genetics, with the chromosome in focus, were originally designated as "cytogenetics" (Liehr and Claussen, 2002). In the 1970s and 1980s, using human chromosomes as opposed to DNA and sequences was extremely common. However, since the beginning of the 1990s, mainstream human and plant geneticists have looked at people dealing with chromosomes as something like outdated "fossils" (Salman *et al.*, 2004). It's interesting that no actual evidence was ever used to justify this attitude. This way of thinking might result from the perception that the more resolution there is, even down to the level of base pairs, as made possible by Sanger sequencing and expedited by recent next-generation sequencing techniques- consequently, genetic studies and diagnostics may become more insightful and significant. Still, it is important to understand that all the yet available techniques to study the genome, at different levels of resolutions, and at the level of the single cell or by examining millions of cells at once, complement rather than compete with one another.

Therefore, cytogenetics as it is discussed here is the study of chromosomes using the classical methods outlined later. It is single-cell oriented, and studies chromosomes for research and diagnostic purposes. It gives researchers the necessary background knowledge regarding chromosome numbers, shapes, and features, such as telomeres, centromeres, chromosome-arms, or banding patterns. The ability to access both mitotic and meiotic chromosomes is also intriguing. In general, cytogenetics serves as the "entry method" when genetic diagnoses are needed for prenatal, postnatal, as well as investigations on karyotypes and genomes in evolutionary study.

One major issue of cytogenetics is the nomenclature of the chromosomes. The latter is not only the basis for description of inborn or acquired chromosomal rearrangements in both plants and humans, but also essential for evolutionary studies of different species. Furthermore, chromosomes are the backbone along which all molecular genetic information is aligned, such as in genomic browsers. Thus, there are international systems for uniform nomenclature guidelines published and regularly updated for humans. To the best of our knowledge there is nothing similar available for plants overall, but there are efforts to provide some standards for heavily studied species like Arabidopsis.

Plant cytogenetics has continued to thrive and contribute significantly to advanced genomic projects by defining contig gaps, determining gene order, and identifying genome rearrangements. Classical plant cytogenetic has now entered the functional genomic era, greatly facilitating the identification, localization, and mapping of chromosome specific markers in plants, which is of high importance in plant breeding, molecular systematics, species identification, detection of hybrid nature, detection of alien chromosomes and chromosomal aberrations, and analysis of somaclonal variations and diversity.

What is Cytogenetics?

"Cytogenetics is essentially a branch of genetics but is also a part of cell biology/cytology that is concerned with how the chromosomes relate to cell behavior, particularly to their behavior during mitosis and meiosis. Techniques used include karyotyping, analysis of G-banded chromosomes and other cytogenetic banding techniques".

A Short History of Cytogenetics

Cytogenetics started with visualization of chromosomes by Walther Flemming in 1879, who also introduced the words "chromatin" and "mitosis," based on his microscopic findings. In 1888, Heinrich Wilhelm Waldeyer introduced the name "chromosome" (meaning "stained body") and Walter Sutton and Theodor Boveri suggested in 1902/1903 the chromosome theory of inheritance. They postulated that Mendel's "hereditary factors" are localized on chromosome pairs. This was further worked out by Thomas Hunt Morgan (1910), who could show that genes are aligned like pearls on a necklace along the chromosome.

The pioneering work on maize by Barbara McClintock had a significant impact on the field of plant cytogenetics (*Zea mays*). Major advancements in our understanding of the structure and dynamic activity of the maize genome were made possible by her method for the precise identification of individual chromosomes (Creighton *et al.*, 1931 and McClintock, 1929, 1932, 1938). McClintock demonstrated using carmine-based chromatin staining techniques that the relative lengths and arm ratios of the chromosomes could be used in combination to uniquely identify each individual chromosome from a single meiotic nucleus. This method worked well for developing cytogenetic maps in additional plant species, such as tomato (*Lycopersicon esculentum*), sorghum (*Sorghum propinquum*), and rice (*Oryza sativa*). However, it was necessary to create further methods for unambiguous cytogenetic resolution and karyotyping in plant species with comparable chromosome sizes.

In 1968, Caspersson *et al.* created additional banding patterns on plant chromosomes called Q-bands using the fluorescent dye quinacrine. Giemsa C-banding and Q-banding on the chromosomes of maize, keeled garlic, and bean (*Vicia faba*) were compared by Vosa and Marchi in 1972. Somatic chromosomes, which are more simpler to obtain than meiotic chromosomes, are much more helpful as a result of the development of chromosome-banding procedures. Individual rice prometaphase chromosomes were identified using Giemsa staining techniques, and diploid rye (*Secale cereale*) and Emir barley (*Hordeum vulgare*). karyotypes were developed In an effort to advance plant cytological methods, Schweizer showed that, with the exception of *Vicia faba*, cold pretreatments improved chromosomal visualisation. Even under ideal staining circumstances, the capacity to differentiate all the inherent physical resemblance of chromosomes in some plant species can undoubtedly cause problems for chromosomes. Variations of the carmine-based and banding techniques have been modified and perfected over many years for cytogenetic characterization of various plant species.

Advantages and Restrictions

Cytogenetics has benefits and shortcomings, as do all approaches. On the plus side:

- ☆ There is a whole genomic view;
- ☆ There is a single-cell view, enabling detection of low-grade mosaics;

- It is a long-standing (experienced since the 1970s), inexpensive approach;
- Aberrations like Robertsonian translocations or heterochromatic supernumerary marker chromosomes can only be picked up by cytogenetics; and
- Cytogenetic data are a prerequisite to interpret correctly high-throughput or complex data on DNA sequences.

On the other hand, there are the following points to consider:

- Resolution is normally not below 5–8 megabases;
- There may be cell cultural artifacts;
- Cell cultivation is time consuming;
- There is a need for trained personnel (however, this is also a problem in other cytogenomic approaches, like next-generation sequencing (Ungelenk, 2021) or molecular karyotyping (Weise and Liehr, 2021); and
- Interphase cells cannot be studied, thus, cytogenetics is restricted to cultivable cell types.

Cytogenetics in Plant Breeding

Cytogenetics can be applied in different phases of plant breeding, and in different ways. Plant breeding as an activity of manipulating genotypes requires techniques and information. Cytogenetics provides both. The information provided can be of a general nature, valid for all materials and for several applications, or it can be specific, relating to the material and/or the problem at hand. The latter can be information on the structure of the starting material, on the material in intermediate stages, and on end products. It can concern the microscopic structure and number of chromosomes (karyotype) and their variation but also the behaviour of chromosomes during transmission, somatic or generative. When special cytogenetic approaches (techniques) in genetic manipulation are used again information is needed, both of a general nature, *e.g.* on the techniques to be employed and of a specific nature on the material to be manipulated. Providing information, therefore, is an essential duty of cytogenetics in plant breeding. Where cytogenetics plays a role in genetic manipulation, it is necessary to distinguish between the manipulation of the genotype (the specific genetic constitution of the genomes under consideration) and manipulation of the genetic transmission or reproductive system. Manipulation of the genotype may involve the introduction or replacement of specific genes, or the manipulation of the gene dose. Manipulation of the transmission system may, for instance, involve the introduction of limitiations to free segregation used in the fixation of heterozygosity.

Table 2.1: The Phases of a Plant Breeding Program (Sybenga 1989)

I	*II*	*III*	*IV*
a) Formulation of objectives	a) Collection of material	a) Selection	b) Propagation/ maintenance
b) Choice of approach in construction of genotype and reproductive system	b) Testing and pre-selection of material	b) Testing	
c) Program design	c) Induction of additional genetic variation (mutation transformation)		
	d) Construction of special reproductive system		
	e) Combination (hybridization and recombination)		

From Cytogenetics to Cytogenomics

Cytogeneticists in particular seem to feel driven by the pressure from molecular geneticists, and have (over) reacted in part by changing well-established designations from "cytogenetics" to "cytogenomics," to appear more "modern," maybe fancier and/or attractive, as the following examples show:

- ☆ For decades, the American Cytogenetic Association has held a conference on chromosomes every 2 years, which was renamed from the "American Cytogenetics Conference" to the "American Cytogenomics Conference" in 2018 (American Cytogenomics Conference, 2020).
- ☆ The well-established "European Cytogenetic Association" also did the same, *i.e.*, their biennial conference, called the "European Cytogenetics Conference" until 2017, has been known as the "European Cytogenomics Conference" only since 2019 (ECA - European Cytogeneticists Association, 2020).
- ☆ The "bible" of human chromosome nomenclature, the ISCN, standing for "international system of cytogenetic nomenclature" since 1978 (Shaffer *et al.*, 2013), was renamed the "international system of cytogenomic nomenclature" in 2016 (McGowan-Jordan *et al.*, 2016).

The designation "cytogenomics" appeared in 1999 and has since then mainly been used as an alternative to "cytogenetics and genomics approaches" (Liehr, 2021). Only in 2019 was it suggested by Russian colleagues that the term cytogenomics could, in retrospect, be deduced as well from the word "cyto (post)genomics" (Iourov, 2019).

A Definition of Cytogenomics

Cytogenomics, as we understand it here, is not at all restricted to diagnostics; such a definition for cytogenomics was given by Silva *et al.* (2019), where it is "a general term that encompasses conventional, as well as molecular cytogenetics (fluorescence in situ hybridization (FISH), microarrays) and molecular-based techniques." Here "cytogenomics" is understood as equivalent wording under the definition used for chromosomics by Prof. Uwe Claussen (Jena, Germany) in 2005 (Claussen, 2005) and outlined in Figure

Prof. Claussen stated that the term "chromosomics," here used as equivalent to the term "cytogenomics," "was introduced to draw attention to the three-dimensional morphological changes in chromosomes that are essential elements in gene regulation" (Claussen, 2005). His idea was to subsume under such a term all chromosome related research with the goal to lead us to novel concepts in biology. This contrasts with other omics-designations, which mainly aim to show the importance of their specific field by separating it from others. Thus, chromosomics/cytogenomics includes the following kinds of studies/fields:

- ☆ Research on plasticity of chromosomes in relation to the three-dimensional positions of genes, which affect cell function in a developmental and tissue specific manner during the cell cycle.
- ☆ Research into chromatin-modification-mediated changes in the architecture of chromosomes, which may influence the functions and lifespans of cells, tissues, organs, and individuals. Getting more insights into the flexible, three dimensional structures of metaphase chromosomes (Daban, 2021) may also help to understand the influence of aforementioned "positional effects" on cells (Weise and Liehr, 2021) at different stages of their development (Pellestor *et al.*, 2021).
- ☆ Research about the occurrence and prevalence of chromosomal gaps and breaks and interchanges, to be studied in the first place by banding cytogenetics (Weise and Liehr, 2021) and molecular cytogenetics (Bisht and Avarello, 2021). Here especially fragile sites and their putative role as seeding points of (i) evolutionary conserved breakpoints, (ii) breakpoints observed in inherited, and (iii) acquired chromosomal aberrations in tumors are of interest. Recently, as suggested by Uwe Claussen, the fragile site-related breaks were attributed to chromosomes' three-dimensional structure and function rather than to DNA-sequence (Liehr, 2019).

Development and Applications of Molecular Cytology in Plants

Regardless of a species' natural chromosome form, the discovery of in situ hybridization (ISH) techniques provided opportunities for cytogenetic assessment of essentially any species (Gall *et al.*, 1969 and Pardue *et al.*, 1969).

	Techniques	Research		Vision
☆	Cytogenetics	Stretching multilayer org.		Novel Concept in biology
☆	Molecular cytogenetics	Interphase architecture		
☆	Molecular karyotyping	Genomic stability CNVs		
☆	Sequencing	Chromothripsis TADs HiC	→	
☆	Optical/karyo-mapping	Repetitive elements		
☆	In situ CRISPr/CAS9	Live cell imaging		
☆	Epigenetic	Long noncoding RNA		

Figure 2.1: Schematic Overview on what Cytogenomics Comprises.

In the upper box the word cytogenomics is written as a pictogram: "C" is depicted like GTG-banded chromosomes; "Y" represents a DNA forming a replication fork; "T" has as *red* upper part standing for a topologically associating domain; "O" is an interphase nucleus, showing the interphase architecture; "G" highlights the epigenetic changes as posttranslational methylation (*green* M) of cytosine; "E" stands for a chromosome, which is micro dissected by a glass needle; "N" includes left hand side a chromosome with one fragile site in the long arm and a so-called gap in the short arm; "O" represents the molecular karyotyping approach; "M" is depicted intentionally in *green* and *red*, representing a rearrangement between to nonhomologous chromosomes stained by two-color fluorescence in-situ hybridization; "I" represents a chromosome in *blue* counter stain DAPI with a *green* and a *red* locus-specific probe as applied in fluorescence in-situ hybridization; "C" includes a pipette, which drops cytogenetically worked-up cell suspension on a slide; "S" is shown as a chromosome 1 after C-banding. In the two left-hand lower boxes some representative approaches/techniques used and actual research applications in cytogenomics are listed; all of them are treated in the following chapters of this book. The right-hand lower box shows that the term cytogenomics is meant to be integrative and include the vision to create novel concepts in biology and medicine and plant breeding.

When creating ISH probes for plants, radioactive tracer or modified nucleotides (attached to biotin, digoxigenin, or fluorescent moieties) allow for microscopic localization and observation of complementary sequences on individual chromosomes as well as in cells and nuclei. According to a recent analysis by Jiang and Gill, direct and indirect fluorescence in situ hybridization (FISH) has been widely used over the past 25 years. Despite the fact that FISH is frequently

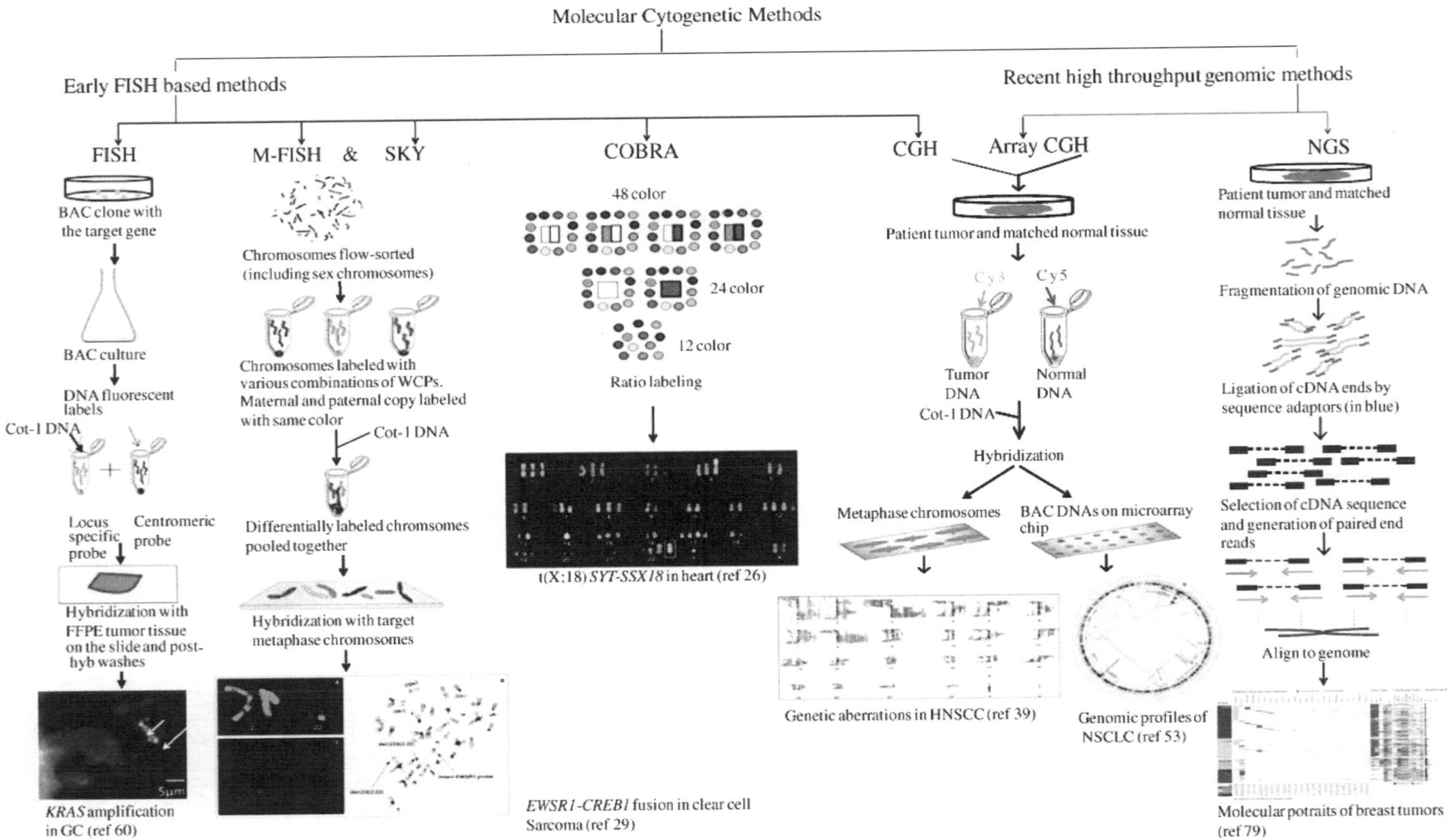

Figure 2.2: Molecular Cytogenetics Method.

used to map distinct or low-copy-number sequences, it is also employed to pinpoint repetitive sequences in order to develop chromosomal recognition cocktails or to investigate genome relationships in polyploid or closely related plant species (Lysak *et al.*, 2001 and Lamb *et al.*, 2006). Plant cytogenetics is in an exceptional position to support, expedite, or direct plant-genome research due to the diverse uses of FISH in structural, comparative, and functional genomics.

FISH (Fluorescence *in situ* hybridization)

A chromosome's particular DNA sequence can be recognized and located in a laboratory environment through a process called fluorescence in situ hybridization (FISH). In plant molecular cytogenetics, FISH is a potent, reliable method for detecting chromosome alterations and determining the distribution of particular DNA sequences. The FISH method was created in 1981 and uses fluorochromes to mark DNA fragments that are then detected by ultraviolet light.

Principle of FISH

Directly in a cell, by chemical treatment and using a higher degree of sequence complementation, probes of DNA or RNA are hybridized on a chromosome(s). The fluorescent-labeled probes when finds their complementary sequence (which is a sequence of our interest), hybridize and emit fluorescence. Signals are recorded and investigated using a fluorescent microscope.

Probe - Molecular level radioactively tagged probes were utilised in the cytogenetic procedures at first because low energy radiation, like the isotope tritium [3H], ensured superior resolution of the probe. Currently, labelling agents consist of non-radioactive DNA probes such biotin and digoxigenin.

Labeling molecules can be found by either direct or indirect staining, such as fluorochromes or an antibody-fluorochrome combination. Fluorescent in situ hybridization is the term given to the procedure when fluorochromes are utilised (FISH).

FISH has shown to be a superb tool for examining the distribution of DNA sequences on chromosomes, copy counts at various sites, and evolutionary changes to the physical structure of the genome.

Types of Target-DNA/Samples in FISH - samples need to contain intact, non-degraded high molecular weight target-DNA.

Accordingly, in multicellular organisms, native cells, extracted nuclei, tissue sections, metaphase chromosomes, or pure DNA can be used as target-DNA Application of FISH ranges from karyotype analysis to gene localisation. In cytogenetics, FISH is used to detect and localise the presence or absence of specific DNA sequences in the chromosomes, and in situ refers to the chromosome as the "original site" for the nuclear DNA to be present. Additionally, Jiang and Gill (2006) utilized this effective method to map repetitive DNA sequences and multi-copy gene families as well as to precisely define the location of genes on chromosomes.

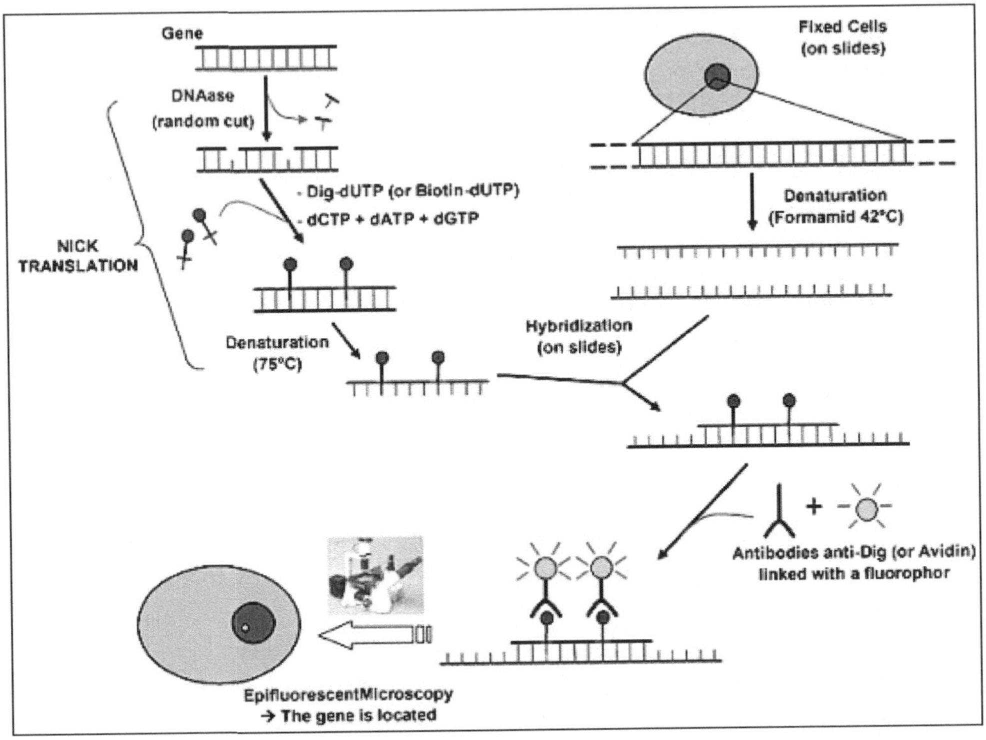

Figure 2.3: Flow Chart of FISH (Fluorescent *In situ* Hybridization).

For the purpose of locating and identifying the locations of cloned DNA sequences on chromosomes, FISH karyotype analysis may be crucial. This study is now a standard technique for locating, positioning, and mapping the genes in several genomes. Cytological preparations of the metaphase chromosome spreads are made from the meristematic root tip cells. Meiotic chromosomes, the preparations of squashed anthers at the early meiosis, prophase or metaphase I, are used. Routinely, the cytological preparations are screened for a sufficient number of cells at a certain stage of division, and the selected spreads are physically adhered to the objective slide by hard freezing.

Use of FISH - FISH allows differential colour labelling of the chromosomes of the parental species in the hybrids or discriminates individual chromosomes in the karyotype, or highlights loci of the specific sequences in a particular chromosome.

FISH to study genome structure in hybrid plants building up physical genome maps and as a tool to study phylogenetic relationships, and there are a few reviews covering these aspects.

It makes use of fluorescent probes, which are nucleic acid sequences that have been tagged and only attach to DNA regions with a high degree of sequence homology. Enzymatic or chemical bonding techniques are used to directly bind fluorescent molecules (FITC, TRITC, Cy 3, Cy 5, *etc.*) to the probe, allowing the probe-target hybrid to be seen under a fluorescence microscope right away after the hybridization reaction. The probe may be made from a cloned DNA sequence, or it may be produced by PCR using flow-sorted chromosomes, microscopically dissected chromosomes, complete genomes, or even mixtures of oligonucleotides. Alphoid centromere sequences (150–250 bp repeats), telomeric/subtelomeric sequences (5–6 bp repeats), unique sequences (100–500 Kb), entire chromosomes, one arm (p or q arm) of a chromosome, or even the entirety of the genomic DNA may be used as probes. Test and probe DNA is annealed following denaturation and visualized by fluorescence microscopy. FISH has been modified to carry out spectral karyotyping/multiplex FISH to indentifying all 24 chromosomes in one experiment and comparative genomic hybridization (CGH) as well as array CGH (aCGH) to screen whole genome.

Types of Probe-DNA Suited for FISH On the one hand, there are commercially available probes, especially for molecular cytogenetics based chromosomic research and diagnostics in humans. These probes are usually ready to use and labeled with corresponding fluorophores or nonfluorescent haptens (for review on commercial probes for cancer cytogenetics (Liehr *et al.*, 2015). The second type of probe-DNA for FISH is in house probes, which need to be labeled, either by PCR-based approaches, Nick-translation, or the so-called Universal Linkage System (ULS) (Liehr, 2017). In the following, five basic types of probe-DNA applied for FISH are listed. Locus Specific Probes Locus-specific probes (LSPs) are normally derived from molecular cloning experiments. Accordingly, genetic vectors, including all kinds of plasmids, bacterial and yeast artificial chromosomes, or others are suited if they contain the wanted insert of species-specific DNA to be targeted by FISH, with inserts of a minimal size of 12 kb (Liehr, 2017). Alternatively, also contiguous probes may be used (Smith *et al.*, 1997), or for mapping purposes, even smaller single copy probes (Nguyen *et al.*, 2019). Repetitive Probes Repetitive DNA can be easily visualized in FISH experiments. Thus, repetitive probes, targeting centromeres, telomeres, or other repetitive, *e.g.*, interspersed satellite-DNAs, result in strong and easily evaluable signals. Interestingly, at least one repetitive DNA (D4Z4) localized in 4q24 has some meaning in human genetic diagnostics and can be traced by molecular combing (Nguyen *et al.*, 2019). Partial Chromosome Paints Partial chromosome paints (pcps) can be established by glass needle-based chromosome microdissection (midi) (Al-Rikabi *et al.*, 2019). Pcps simultaneously stain at least 1 or 2 euchromatic chromosomal subbands and are normally not larger than a chromosome arm. Whole Chromosome Paints A whole chromosome paint (wcp), staining an entire chromosome can either be established by midi (Ferguson Smith *et al.*, 2005) or by chromosome flow sorting.

Table 2.2: Important Achievements of using FISH Techniques in Cytogenomics, Mapping and Genome Analysis of some Major Food Crops

Crop	*Technique*	*Achievement*
Cereals (rice, wheat, maize, rye, barley, oat, sorghum)	M-FISH, I-FISH, Fiber-FISH, BAC-FISH, YAC-FISH, Pachytene-FISH, ultrasensitive FISH, combined FISH with Immuno techniques, extended DNA fiber, Super stretched pachytene chromosome	***Rice****, localization of glutelin gene, gall* midge resistance gene (*Gm2*) *on* chromosome 1, bacterial blight resistance locus *Xa21 and blast resistance gene Pi-b* on chromosome 2, ***Maize:*** *painting of all somatic* chromosomes, localization of centromere in relation to genetically mapped markers, characterization of centromere- specific histone H3 variant, immune detection of DNA methylation; ***Barley:*** *identification* and polymorphism of *Hordeum chilense* chromosomes and detection of BARE 1 retrotransposon; ***Oat:*** *karyotyping of Avena* species, phylogenomic analysis; new chromosome nomenclature from *Avena*monosomic line; ***Sorghum:*** *molecular* cytogenetic map, *sh2 gene, liguleless* linkage group in sorghum; ***Rye:*** *analysis of* B chromosomes, telomeric heterochromatin, detection of rare translocation, activity of rRNA genes, identification of chromosome complement in hybrid.
Grain legumes (pea, chickpea, grasspea, beans, lentil)	Dual-color FISH, BAC-FISH,	Quantitative karyotyping, identification of NOR (nucleolus organizer region), mapping of 5S and 18S–25S rDNA probes, rDNA evolution, amplification of rDNA, detection of aneuploids (trisomics), analysis of B4 resistance (R) gene cluster in common bean and macrosynteny with *Medicago truncatula and Lotus japonicas* in chromosomes *Mt6 and Lj2, Rdna* localization in chickpea, karyotyping of lentil chromosomes, phylogenetic relationship, location of chromosome telomeres, detection of NOR and satellite DNA sequence families in different *Lathyrus, Pisum, and Cicer species*
Oil-yielding (Brassica, soybean, peanut, sunflower)	DNA combing-FISH, BAC-BIBAC-FISH, BAC-SSR-FISH	***Brassica****: a 76-kb fragment in a P1-derived* artificial chromosome (PAC) clone containing the *SLG and SRK* (*self incompatibility* locus) genes was used to directly visualize the *S locus. Using DNA* combing and FISH, the positions of the Fluorescent signals of *SLG and SRK on the* clone are found consistent with their positions on the restriction map; chromosomal rearrangement through homoeologous recombination in *B. napus,* detection of amphidiploids and chromosome triplication across Brassicaceae; ***Sunflower:*** *rRNA mapping,* mapping of desired agronomic trait, detection of single copy sequence, construction of molecular cytogenetic map;

Crop	Technique	Achievement
		Soybean: *molecular karyotyping,* translocation mapping, detection of segmental duplication, validation of mutagenesis and TILLING; ***Peanut:*** confirmation of *Arachis duranensis and A. ipaensis as the wild diploid progenitors* of *A. hypogaea, evidence for origin of* autotriploid *A. pintoi.*
Vegetables (potato, tomato, cucumber)	FISH, cross species, BAC-FISH, multi color FISH, pooled BAC-FISH, DAPI-FISH, SC-FISH, RAPD-FISH	Karyotyping, DAPI and rDNA mapping, genome sequencing, high-resolution chromosome mapping, detection of chromosomal rearrangement, assignment of genetic linkage map to pachytenechromosome, quantification of euchromatin and heterochromatin, localization of single or low-copy sequences on **tomato** chromosome, high-resolution karyotyping, mapping of 45SrRNA loci, construction of integrated molecular cytogenetic map in **cucumber,** synteny between *Cucumis sativus and C. melo and comparative* karyotyping with fosmid
Horticultural crops	FISH, Fiber-FISH, multicolor FISH, pachytene-FISH	Variation and mapping of rRNA gene in ***Lycoris spp***. (*Amaryllidaceae*), *Paphiopedilum and Dendrobium* (Orchidaceae), mapping of 5S and 45S rRNA and species relationship in *Tagetes, Aster,* ***Chrysanthemum*** (*Asteraceae*), karyotyping using centromeric repeat in ***Antirrhinum majus*** (*Scrophulariaceae*), *Hibiscus* (*Malvaceae*), ***Rosa spp.*** (*Rosaceae*)

GISH (Genomic *in situ* hybridization)

Numerous modifications have been made to the DNA sequences in situ hybridization method since it was initially used. One of these modifications is the genomic in situ hybridization (GISH) method, in which the probe is made of the genomic DNA of another organism and the targeted DNA of the target organism. (Valente *et al.,* 2009).

GISH is an advanced technique that can distinguish between parental genomes, display the genome organisation of an interspecific hybrid, and as a result, locate (promising) recombination sites in the chromosomes of allopolyploid and interspecific introgression lines (Jiang *et al,.* 1995). In contrast to FISH, which uses a specific region of DNA as a probe, the GISH approach uses the entire genome of a species (Xie, 2012). In GISH analysis, more DNA is available, and therefore, specific probes are easier to identify and amplify (Pen˜aloza and Pozzobon, 2007). However, the ratio of blocking DNA or probe DNA should be sufficient to prevent both genomes from being simultaneously labelled with chromosomes (Brammer *et al.,* 2009). Because there is a considerable likelihood that homology will manifest during hybrid generation in hybrids generated from closely related species, blocking DNA plays an important role in these cases (Parokonny *et al.*, 1997; Xie *et al.*, 2010).

This widely used cytogenetic technique provides a direct visual examination for separating the parental genomes and can be used to look into interspecific introgression lines, allopolyploid species, and hybrids (Jiang and Gill 1994). Finding alien chromatin stages and integration positions can also be done with the use of this instrument in a precise and effective manner. Additionally, GISH has been effectively applied in recent years to identify a number of intergenomic, intersectional, and interspecific Lilium hybrid cultivars (Lim *et al.*, 2000, 2001, 2003; Zhou *et al.*, 2008; Khan *et al.*, 2010). The GISH method also provides a fresh tool for a more accurate parental genome analysis in somatic and sexual hybrids. This methodology allows for the investigation of painted chromosomes during the whole cell cycle as well as the identification of translocation-related chromosomes from different genomes (Kenton *et al.*, 1993; Zhao *et al.*, 2013).

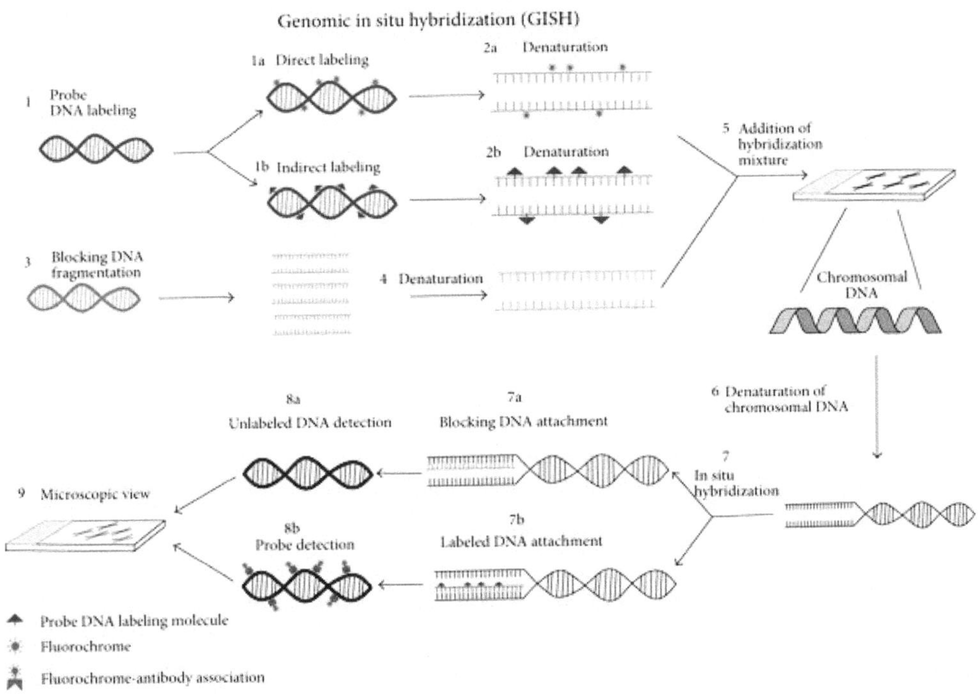

Figure 23: Flow Chart of GISH (Genomic In situ Hybridization).

Use of GISH

Meiotic behaviour, including chromosome behaviour during the meiotic cycle, the chromosome pairing process, the behaviour of artificial and natural hybrids, and homologous and homeologous chromosome pairing, can be studied using GISH (Xie *et al.*, 2010). By linking them to the genitor genomes, the GISH approach may also measure the multivalent, bivalent, and univalent formation rates in hybrids as well as the frequency of chiasma junctions between homologous chromosomes. Another fundamental advantage is that this method allows us to identify the causes that lead to abnormal meiosis and how they may affect fertility (Silva and Souza, 2013). The presence of hybrids formed from apomictic species, which can procreate without fertilisation, is further confirmed by GISH. Some species of the genus Lilium have already shown evidence of apomixis. Therefore, it is crucial to determine if the hybrids produced by the union of these species are real hybrids or not. GISH provides a clear identification of this. GISH can be used to confirm allopolyploid species and hybrid plants because there is a distinct and unmistakable difference between the genomes. This technique can be used to determine the genomic origin during prophase and metaphase chromosome spreads, as well as in interphase nuclei, where there is no prior knowledge of chromosomal morphology (Marasek *et al.*, 2004). The GISH method is the only one that can pinpoint intergenomic recombination, genome makeup, and the source of polyploid hybrids and their offspring. (Ali *et al*,. 2004; Chung *et al.*, 2013).

CGH (Comparative Genomic Hybridization)

With the help of the CGH FISH technique, it is possible to perform a thorough analysis of chromosomal imbalance (relative copy number as gains or losses) in the entire genome (molecular karyotyping at cytogenetic resolution) in a single test without the need for metaphase preparation or cell culture of the test samples. Similar to dual FISH, the CGH technique relies on co-hybridizing differentially labelled whole genomic test and control DNA in an equal ratio on normal metaphase spread. The slide is seen using an epi fluorescence microscope following the usual washing procedure. Software is used to process the images and quantify the fluorescence intensities over the whole length of each chromosome that has been karyotyped. All chromosomes are stained generally as a result of hybridization. Hybridization will show more test DNA (or more green, if the test DNA is coloured with green) at the relevant target region of the hybridised chromosome if the test DNA includes additional copies of the DNA material. Similar to deletion, monosomy will result in weaker signal levels. However, balanced translation, polyploidy/tetraploidy, and mosaicism of less than 50 per cent cannot be detected by CGH. Although straightforward and inexpensive to produce, CGH requires a lot of work, takes a long time, and is challenging to interpret (because karyotyping of DAPI banded chromosomes is erroneous), hence array CGH has taken its place.

With a few modifications, array CGH is carried out using the same basic principles as CGH (co-hybridization is done on DNA spots or arrays (BAC/PAC/oligonucleotides) rather than metaphase chromosomes). The slide is scanned to determine the fluorescence intensities at each target on the microarray following hybridization of differently labelled test and reference genomic DNA to the target sequences. The position of the sequence along the chromosomes is then shown against the normalised fluorescence ratio for the test and reference DNA (as in CGH). Values higher or lower than the typical 1:1 ratio indicate gains or losses throughout the genome. The size, quantity, and positioning of the targets on the genome all affect the resolution. The full genome can be viewed at once, and the technique is quick (24–36 hours). This approach involves no live cells or cell culture, is rapid (24–36 hours), and delivers a full genome view at an extremely high resolution. Without any prior knowledge of problems, it may detect the majority of microscopic and sub-microscopic chromosomal alterations from any DNA source in a single experiment. The sole drawback is that, unlike CGH, it cannot detect polyploidy or balanced chromosome aberrations. It is currently suggested as a first line test by a number of authority for assessing numerous abnormalities, developmental/mental retardation, prenatal diagnosis, *etc.* The field of cancer, which offers aCGH great potential, is another significant application. SNP microarray is an advancement of aCGH that provides improved resolution in addition to the ability to detect mosaicism, paternal origin, zygosity, translocation, and other traits.

Multiplex in situ hybridization (M-FISH), first described by Speicher (1) enables the simultaneous visualization of all chromosomes in a single hybridization. The technique, which essentially paints the entire genome in multiple colors, was developed in the same year as the related SKY technique (2) and represented a major achievement in molecular cytogenetics. Technologically, both M-FISH and SKY combined the significant advances that had been made in fluorescent probe labeling strategies (3), digital imaging fluorescence microscopy, and image processing capability, to create revolutionary karyotypic tools for the analysis of structural and numerical abnormalities (4). The power of M-FISH (and SKY) lies in its ability to resolve complex karyotypes and identify the origin of marker chromosomes as evidenced by its many applications in tumor diagnostics and research, for example (5, 6), evolutionary cytogenetics (7) and in the study of chemical and radiation-induced aberrations (8). Impressively, even quite complex karyotypes of individual cells within a nonclonal population can be resolved with a high degree of confidence.

A multiplex probe set for M-FISH consists of whole chromosome probes (WCP) each labeled with one of a finite number of spectrally distinct fluorophores. For example, the majority of commercially available human probe sets (*e.g.*, human M-FISH (Genetix) or 24 X-Cyte (Metasystems) consist of 52 WCP directly and indirectly labeled with one of five fluorophores such that each chromosome is uniquely labeled entirely down its whole length with one, two, or three different fluors. It is this combinatorial labeling strategy of fluorophores that forms the basis of M-FISH chromosome classification.

Use - M-FISH is a very powerful technique for detecting interchromosomal alterations (~3–10 Mb in size) (12) of varying complexity, throughout the genome. At its most simplistic, structural rearrangements are identified as discontinuities and changes in M-FISH pseudocolor down the length of DAPI counterstained chromosomes. Theoretically therefore, it is possible to analyze each metaphase using just the M-FISH color karyotype alone. However, this is not recommended for a number of reasons. Firstly, chromosome classification is determined on a pixel by pixel basis and is therefore very sensitive to subtle changes in the signal: noise ratio of component fluors across and down the length of individual chromosomes.

Advantages of aCGH Technology

One assay utilising this technology is equivalent to thousands of FISH assays, saving time and money. The main benefit of aCGH is the capacity to concurrently detect aneuploidies, deletions, duplications, and/or amplifications of every locus represented on an array. Additionally, aCGH has demonstrated to be an effective method for identifying submicroscopic chromosomal abnormalities in people with idiopathic mental retardation and other birth disorders. In fact, aCGH has a 10 per cent–20 per cent detection rate of chromosomal abnormalities in children with mental retardation/developmental delay with or without congenital defects; only 3 per cent –5 per cent of these abnormalities would be detectable by conventional methods, according to a number of large-scale investigations. For instance, 1,049 (11.9 per cent) of 8,789 patients evaluated by aCGH had a clinically significant mutation relevant chromosomal abnormality.

CGH Technique

1. **Preparation of Metaphase slides:** chromosomes in metaphase are used as the target DNA. Colchicine is then used to arrest cells in mitosis and the cells are then dropped onto slide in such a way that they are spread nicely and don't overlap.
2. **Extraction of Test and Reference DNA:** The test DNA is extracted from fresh/frozen bulk tissue or paraffin embedded tissue.
3. **Labelling and Fragmentation Of Test and Reference DNA:** Is carried out by a process called Nick Trans lation where the Test and Reference DNA are labelled directly using different Fluorochromes -Test DNA produces a Green Fluorescence -Reference DNA produces a Red Fluorescence.
4. **Hybridisation:** Equal amounts of the Test and Reference DNA labelled DNA compete to undergo complementary base pairing with specific sequences on metaphase spreads.
5. **Use Of Fluorescence Microscopy and Digital Image Analysis:** The fluoresce ratio of Red – Green intensity is calculated in order to determine whether chromosomal Loss or Gain has occurred.

Fibre-FISH

During the past few years, fluorescence in situ hybridization to DNA fibers (Fiber–FISH) (Raap *et al.*, 1996) has gained considerable interest in genomics as

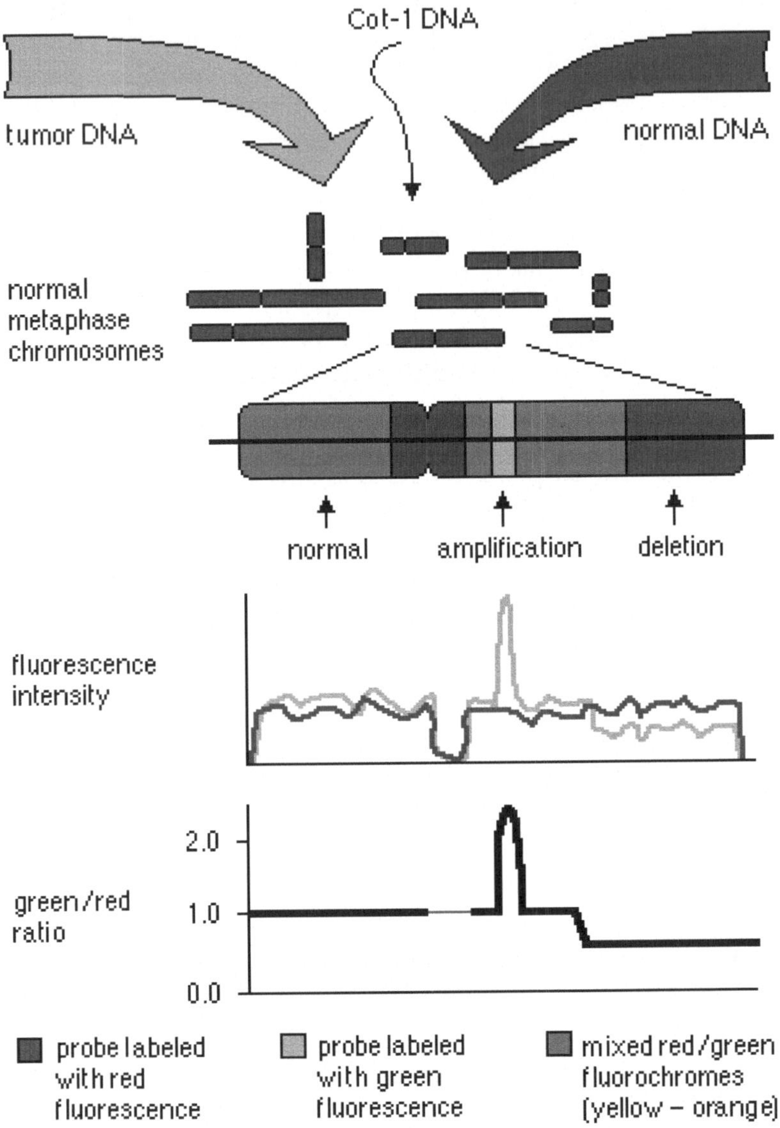

an adjunct physical mapping tool (Heiskanen *et al.*, 1995) It also proved to be ideally suited for molecular pathology research due to its high and wide-ranging resolution (1 kb-1 Mbp) and gene colour bar-coding capability (Florijn *et al.*, 1995). This is because, in comparison to Southern hybridization- or polymerase chain reaction-based study, it offers a substantially larger genomic scope for the investigation of DNA rearrangements in hereditary disorders. (Michalet *et al.*, 1997) As hybridization probes in Fiber-FISH, big insert clones like cosmids, bacterial, and yeast artificial chromosomes are frequently utilised. Specific hybridization to whole genomes or cloned DNA targets is identified as linear or nearly linear lengths of beads-on-a-string fluorescence signals. The usual discontinuous Fiber-FISH pattern has been attributed to a number of factors, including suppression of repeat sequences, loss and inaccessibility of target DNA due to either in situ renaturation or adhesion to the glass substrate (Windle *et al.*, 1995). However, there is a dearth of experimental evaluation of the variables affecting the pattern's nature. Here, we present results from research that sought to better understand the cause of the beads-on-a-string characteristic of Fiber-FISH signals by tracking the accessibility and loss of DNA in a cloned target during the FISH process. In contrast to nick-translation or random primed labelling, the chemical nucleic acid labelling method we adopted for this purpose leaves the sugar-phosphate backbone intact (Wiegant *et al.*, 1999). Thus, intact labeled target DNA molecules can be produced for monitoring their fate during Fiber–FISH.

Amplified or single molecule templates are used in NGS/massive parallel sequencing technology for massively parallel sequencing. Throughput is increased by several orders of magnitude as a result. Targeted gene, exome, and genome analyses are the three main levels of sequencing analysis. It can identify practically all major types of genomic modifications, such as balanced translocations, nucleotide substitutions, minor insertions and deletions, copy number variations, mosaicism, parental inheritance, and the integration of foreign or microbial DNA. Because this method does not require cell culture, has a subjective interpretation process, and provides information more quickly than conventional cytogenetics, it is posing a serious threat to conventional cytogenetics (balanced translocation detection advantage of conventional cytogenetics is also addressed by NGS). As a result, NGS are now being widely used in clinical settings.

This century is observing a paradigm shift from cytogenetics to cytogenomics. NGS and SNP microarray based genome analysis are becoming acceptable tools for research and patient care. We are going to witness gradual fall of conventional cytogenetics usages in near future.

Conclusion

Cytogenetics is a branch of genetics that studies the structure and function of chromosomes in cells. It has become increasingly relevant in modern breeding efforts, as understanding the chromosome composition of plants and animals is crucial for identifying desirable traits and developing new varieties with improved characteristics. For example, in plant breeding, cytogenetics is used to create hybrid varieties with higher yields, resistance to pests and diseases, and improved quality of fruits and vegetables. Cytogenetics is also used in the development of new crops with increased nutritional value and improved tolerance to environmental stresses such as drought and high temperatures. In animal breeding, cytogenetics plays an important role in improving livestock production, such as increasing milk and meat production, improving resistance to diseases, and reducing the rate of inherited disorders. Additionally, cytogenetics is also used in medical research, where it plays a crucial role in the diagnosis and treatment of genetic disorders. By studying the chromosomes of patients with genetic diseases, scientists can identify the specific genetic mutations responsible for the condition, which can lead to the development of new diagnostic and therapeutic approaches. Overall, the study of cytogenetics has broad applications in modern agriculture and medicine, and is essential for the advancement of breeding programs and the development of new treatments for genetic disorders.

REFERENCES

Ali H, Lysak M, Schubert I. Genomic in situ hybridization in plants with small genomes is feasible and elucidates the chromosomal parentage in interspecific Arabidops. 2004.

Al-Rikabi ABH, Cioffi MB, Liehr T. Chromosome Microdissection on Semi-archived Material. *Cytometry A*. 2019; 95, 1285–1288.

American Cytogenomics Conference. 2020; https: //chromophile.org/.

Bisht P and Avarello MDM. Molecular combing solutions to characterize replication kinetics and genome rearrangements. In T. Liehr (Ed.), *Cytogenomics*. 2021; 47–71.

Brammer SP, Poersch LB, Oliveira AR, Vasconcelos S. Hibridização Genômica in situ em Triticeae: Um Enfoque Metodológico. Embrapa Trigo, *Passo Fundo*. 2009.

Caspersson T, Farber S, Foley GE, *et al.*, Chemical differentiation along metaphase chromosomes. *Exp Cell Res*. 1968; 49: 219–22.

Chung MY, Chung JD, Ramanna M, van Tuyl JM, Lim KB. Production of polyploids and unreduced gametes in Lilium auratum × L. henryi hybrid. *Int J Biol Sci*. 2013; 9: 693–701.

Claussen U. Chromosomics. *Cytogenetic and Genome Research*. 2019; *111*(2), 101–106.

Creighton HB, McClintock B. A correlation of cytological and genetical crossing-over in *Zea mays*. *Proc Natl Acad Sci USA*. 1931; 17: 492–7.

Daban JR. Multilayer organization of chromosomes. In T. Liehr (Ed.), *Cytogenomics*. 2021; 267–296.

ECA—European Cytogeneticists Association. 2020; https: //www.e-c-a.eu/en/.

Ferguson-Smith MA, Yang F, Rens W, O'Brien PCM. The Impact of Chromosome Sorting and Painting on the Comparative Analysis of Primate Genomes. *Cytogenet. Genome Res*. 2005; 108, 112–121.

Florijn RJ, Blonden LA, Vrolijk H, Wiegant J, Vaandrager JW, Baas F, den Dunnen JT, Tanke HJ, van Ommen GJB, Raap AK. High-resolution DNA Fiber-FISH for genomic DNA mapping and colour bar-coding of large genes. *Hum Mol Genet.*1995; 4: 831–836.

Gall JG, Pardue ML. Formation and detection of RNA DNA hybrid molecules in cytological preparations. *Proc Natl Acad Sci USA*. 1969; 63: 378–83.

Heiskanen M, Hellsten E, Kallioniemi OP, Makela TP, Alitalo K, Peltonen L, Palotie A. Visual mapping by fiber-FISH. *Genomics*. 1995; 30: 31–36.

Iourov IY. Cytopostgenomics: What is it and how does it work? *Current Genomics*. 2019; *20*(2), 77–78.

Jiang J, Gill BS. Current status and the future of fluorescence in situ hybridisation (FISH) in plant genome research. *Genome*. 2006; 49 (9): 1057–1068.

Jiang JB, Gill S, Wang GL, Ronald PC, Ward DC. Metaphase and interphase fluorescence in situ hybridization mapping of the rice genome with bacterial artificial chromosomes. *Proc Natl Acad Sci USA*. 1995; 92: 4487–4491

Jiang JM, Gill BS. Non-isotopic in situ hybridization and plant genome mapping: the first 10 years. *Genome*. 1994; 37: 717–725.

Kenton A, Parokonny AS, Gleba YY, Bennett MD. Characterization of the Nicotiana tabacum L. genome by molecular cytogenetics. *Mol Gen Genet*. 1993; 240: 159–169.

Khan N, Barba-Gonzalez R, Ramanna MS, Arens P, Visser RGF, van Tuyl JM. Relevance of unilateral and bilateral sexual polyploidization in relation to intergenomic recombination and introgression in Lilium species hybrids. *Euphytica*. 2010; 171: 157–173.

Lamb JC, Birchler JA. Retroelement genome painting: cytological visualization of retroelement expansions in the genera *Zea* and *Tripsacum*. *Genetics* 2006; 173: 1007–21.

Liehr T and Claussen U. Current developments in human molecular cytogenetic techniques. *Current Molecular Medicine*. 2002; 2(3), 283–297.

Liehr T, Othman MA, Rittscher K, Alhourani E. The Current State of Molecular Cytogenetics in Cancer Diagnosis. *Expert Rev. Mol. Diagn*. 2015; 15, 517–526.

Liehr T. Fluorescence in Situ Hybridization (FISH) – Application Guide. 2nd ed. Berlin., Germany. *Springer*. 2017.

Liehr T. From human cytogenetics to human chromosomics. *International Journal of Molecular Sciences*. 2019; *20*, 826.

Liehr T. Overview of currently available approaches used in cytogenomics. In T. Liehr (Ed.), *Cytogenomics*. 2016; 11–24.

Lim KB, Chung JD, Van Kronenburg BCE, Ramanna MS, De Jong JH, Van Tuyl JM. Introgeression of Lilium rubellum Baker chromosomes into L.

longiflorum Thunb.: a genome painting study of the F1 hybrid, BC1 and BC2 progenies. *Chromosome Res.* 2000; 8: 119–125.

Lim KB, Ramanna MS, De Jong JH, Jacobsen E, Van Tuyl JM. Indeterminate meiotic restitution (IMR): a novel type of meiotic nuclear restitution mechanism detected in interspecific lily hybrids by GISH. *Theor Appl Genet.* 2001; 103: 219–230.

Lim KB, Ramanna MS, Jacobsen E, Van Tuyl JM. Evaluation of BC2 progenies derived from 3x–2x and 3x–4x crosses of Lilium hybrids: a GISH analysis. *Theor Appl Genet.* 2003; 106: 568–574.

Lysak MA, Fransz PF, Ali HBM, *et al.*, Chromosome painting in Arabidopsis thaliana. *PlantJ.* 2001; 28: 689–97.

Marasek A, Hasterok R, Wiejacha K, Orlikowska T. Determination by GISH and FISH of hybrid status in Lilium. *Hereditas.*2004; 140: 1–7.

McClintock B. Chromosome morphology in *Zea mays. Science.* 1929; 69: 629.

McClintock B. The production of homozygous deficient tissues with mutant characteristics by means of the aberrant mitotic behavior of ring-shaped chromosomes. *Genetics.* 1938; 23: 315–76.

McClintock B. The stability of broken ends of chromosomes in *Zea mays. Genetics.* 1932; 26: 234–82.

McGowan-Jordan J, Simons J, and Schmid M. (Eds.). *ISCN 2016—An international system for human cytogenomic nomenclature.* 2016.

Michalet X, Ekong R, Fougerousse F, Rousseaux S, Schurra C, Hornigold N, Slegtenhorst Mv, Wolfe J, Povey S, Beckman JS, Bensimon A. Dynamic molecular combing: stretching the whole human genome for high resolution studies. *Science.* 1997; 277: 1518–1523.

Nguyen K, Broucqsault N, Chaix C, Roche S, Robin JD, Vovan C, *et al.*, Deciphering the Complexity of the 4q and 10q Subtelomeres by Molecular Combing in Healthy Individuals and Patients with Facioscapulohumeral Dystrophy. J. Med. *Genet.* 2019; 56, 590–601.

Pardue ML, Gall JG. Molecular hybridization of radioactive DNA to the DNA of cytological preparations. *Proc Natl Acad Sci USA.* 1969; 64: 600–4.

Parokonny AS, Marshall JA, Bennett MD, Cocking EC. Homoeologous paring and recombination in backcross derivatives of tomato somatic hybrids [Lycopersicon esculentum (+) L. peruvianum]. *Theor Appl Genet.* 1997; 94: 713–723.

Pellestor F, Gaillard JB, Schneider A, Puechberty J, and Gatinois V. Chromoanagenesis phenomena and their formation mechanisms. In T. Liehr (Ed.), *Cytogenomics.* 2021; 213–245.

Peñaloza APS, Pozzobon MT. Caracterização citogenética de germoplasma vegetal. In: Nass LL (ed) Recursos genéticos vegetais. Embrapa Recursos *Genéticos e Biotecnologia, Brasília*. 2007; 308–342.

Raap AK, Florijn RJ, Blonden LAJ, Wiegant J, Vaandrager JW, Vrolijk H, den Dunnen JT, Tanke HJ, van Ommen GJB. Fiber FISH as a DNA mapping tool. Methods. 1996; 9: 67–73.

Salman M, Jhanwar SC, and Ostrer H. Will the new cytogenetics replace the old cytogenetics? *Clinical Genetics*. 2004; 66(4), 265–275.

Schweizer D. Differential staining of plant chromosomes. *Chromosoma*. 1973; 40: 307–20.

Shaffer L, McGowan-Jordan J, and Schmid M. (Eds.). *ISCN 2013—An international system for human cytogenetic nomenclature*. 2013.

Silva GS and Souza MM. Genomic in situ hybridization in plants. *Genet Mol Res*. 2010; 12: 2953–2965.

Silva M, de Leeuw N, Mann K, Schuring-Blom H, Morgan S, Giardino D, Hastings R. European guidelines for constitutional cytogenomic analysis. *European Journal of Human Genetics*. 2019; 27(1), 1–16.

Smith CM, Ma NS, Nowak NJ, Shows TB, Gerhard DS. A 3-Mb Contig fromD11S987toMLK3,a Gene-Rich Region in 11q13: *Genome Res*. 1997; 7, 835–842.

Ungelenk M. Sequencing approaches. In T. Liehr (Ed.), *Cytogenomics*.2021; 87–122.

Valente GT, Carlos HS, Maria CG, Eliana F, Cesar M. Comparative cytogenetics of cichlid fishes through genomic in situ hybridization (GISH) with emphasis on Oreochromis niloticus. *Chromosome Res*. 2009; 17: 791–799

Vosa CG, Marchi P. Quinacrine. fluorescence and Giemsa staining in plants. *Nat New Biol*. 1972; 237: 191–2.

Weise A, and Liehr T. Molecular karyotyping. In T. Liehr (Ed.), *Cytogenomics*. 2021; 73–85.

Wiegant J, van Gijlswijk RPM, Heetebrij RJ, Bezrookove V, Raap AK, Tanke HJ. ULS: a versatile method of labeling nucleic acids for FISH based on a monofunctional reaction of cisplatin derivatives with guanine moieties. Cytogenet Cell Genet.1995; 87: 47–52.

Windle B, Silvas E, Parra I. High resolution microscopic mapping of DNA using multi-color fluorescent hybridization. *Electrophoresis*. 1995; 16: 273–278.

Xie S, Khan N, Ramanna MS, Niu L, Marasek-Ciolakowska A, Arens P, van Tuyl JM. An assessment of chromosomal rearrangements in neopolyploids of Lilium hybrids. *Genome*. 2010; 53: 439–446.

Xie S. A molecular cytogenetic analysis of chromosome behavior in Lillium hybrids, PhD-thesis. Wageningen University, Wageningen, *The Netherlands*. 2012.

Zhao HB, Zhao LJ, Zhao LW, Yuan HM, Zhang YM, Xin WL, Sun LF, Guo CH, Li JL. Chromosome deletion and translocation in chinese spring-lophopyrum elongatum 6E disomic addition lines induced by the gametocidal chromosome 2C from Aegilops cylindrical. *Biotechnol Biotechnol Equip*. 2010; 27: 3638–3643.

Zhou S, Ramanna MS, Visser RGF, van Tuyl JM. Analysis of the meiosis in the F1 hybrids of Longiflorum × Asiatic (LA) of lilies (Lilium) using genomic in situ hybridization. *J Genet Genomics*. 2008; 35: 687–695.

Chapter 3

Recent Updates on Chromosomal Aberrations

Himakara Datta M., Sampath L., Tushar Arun Mohanty and B.J. Antony*

Ph.D. Scholar (GPB), Centre for Genetics and Plant Breeding and Genetics, Tamil Nadu Agricultural University, Chennai, Tamil Nadu
****e-mail: mhdatta@gmail.com***

Introduction

A chromosome is a long, coiled DNA molecule that contains genetic information in the form of genes. Chromosomes exist in the nucleus of eukaryotic cells and are typically organized in pairs. The number and type of chromosomes can vary between species. Chromosomes play a crucial role in determining an organism's physical and behavioral characteristics, as well as its ability to pass those traits down to future generations.

The structure of a chromosome can be divided into several parts, such as the chromatin which makes up chromosomes, and is composed of DNA and proteins, including histones, the centromere, the region where the two sister chromatids are held together, telomeres which are the repetitive DNA sequences found at the ends of chromosomes that protect the genetic information. The regions of the chromosome located on either side of the centromere are called the arms.

The main functions of chromosomes are:

- ☆ Storage of genetic information
- ☆ Facilitation of cell division
- ☆ Regulation of gene expression
- ☆ Evolution by carrying mutations and variations in the genetic code, which can be passed on to future generations and potentially result in the development of new species over time.

Plant chromosomes are similar in structure and function to chromosomes in other organisms, but there are also some unique features that are specific to plants. Some of the key characteristics of plant chromosomes include:

- ☆ Number: The number of chromosomes in a plant cell varies between species, but can range from a few to several hundred.
- ☆ Polyploidy: Many plant species are polyploidy, meaning they have multiple sets of chromosomes. This can increase genetic diversity and provide a survival advantage in certain environments.
- ☆ Size: Plant chromosomes are generally larger in size compared to animal chromosomes, which can make them more difficult to study and manipulate.
- ☆ Chromosome structure: Plant chromosomes are usually more complex in structure than animal chromosomes, with a higher degree of folding and compaction of the DNA molecule.

Chromosome behavior during cell division: Plant cells undergo a unique type of cell division called mitosis, which allows the chromosomes to be duplicated and separated into daughter cells with a complete set of genetic information.

Overall, plant chromosomes play a vital role in the growth, development, and reproduction of plants, as well as in the adaptation to changing environmental conditions.

Structure of Chromosomes

A chromosome is a complete chain of DNA and associated stabilizing proteins, appearing as thread-like structures within the nucleus of both animal and plant cells. The DNA double helix is condensed into compact chromosomes during certain stages of the cell cycle by wrapping short segments of the DNA around a core of eight histones at regular intervals. This combination of DNA and histones is referred to as chromatin. The resulting bead-like structure composed of histones and DNA is called a nucleosome, while the DNA that connects the nucleosomes is called linker DNA. The nucleosome compacts the DNA so that it is seven times shorter than the length of the double helix alone and has a diameter of 10 nm, compared to the 2-nm diameter of the DNA

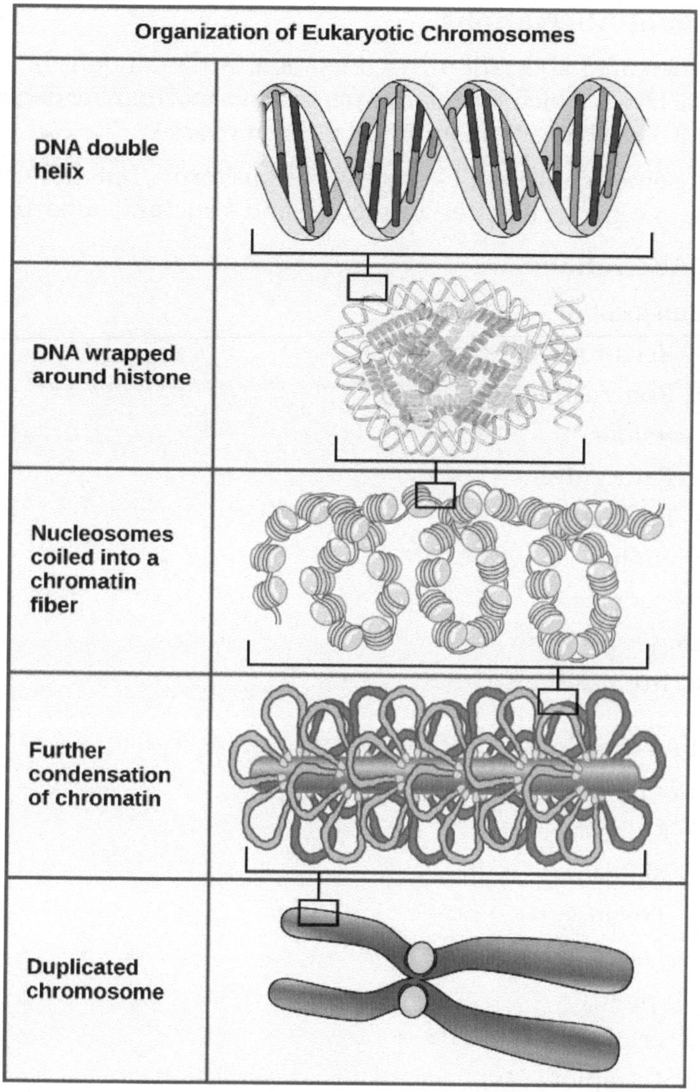

Figure 3.1: Packaging of DNA into Chromosomes (*Source*: Lumen learning).

double helix. The nucleosomes and linker DNA are then coiled to form a 30-nm chromatin fiber, further shortening the chromosome and reducing it to about 50 times shorter than its extended form. The final level of compaction is achieved through the use of fibrous proteins that pack the chromatin, ensuring that each chromosome has its own designated area within the nucleus of a non-dividing cell, without overlapping with other chromosomes. A chromosome's type can be determined by the location of its centromere, resulting in four different categories: telocentric, acrocentric, sub-metacentric, and metacentric.

Chromosomal Aberrations

A chromosomal aberration is a deviation in the amount or structure of chromosomal DNA, which can manifest as an abnormal number of chromosomes or a change in the shape of an individual chromosome.

Chromosome abnormalities come in various forms, but they can generally be classified into two categories: numerical and structural abnormalities.

Structural Aberrations

- ✰ Translocation
  - ❖ Reciprocal
  - ❖ Robertsonian
- ✰ Inversions
  - ❖ Paracentric
  - ❖ Pericentric
- ✰ Deletions
- ✰ Insertions
- ✰ Rings
- ✰ Isochromosomes

Numerical Aberrations

- ✰ Aneuploidy
  - ❖ Monosomy (n-1)
  - ❖ Nullisomy (n-2)
  - ❖ Trisomy (n+1)
  - ❖ Tetrasomy (n+2)
- ✰ Polyploidy
  - ❖ Triploid (3x)
  - ❖ Tetraploid (4x)
  - ❖ Hexaploid (6x)
  - ❖ Octaploid (8x)

Chromosomal Aberrations and their Cause

Structural chromosomal aberrations refer to changes in the structure of chromosomes, including deletions, duplications, inversions, and translocations. These changes can affect the normal functioning of genes and cause a variety of genetic disorders. They can occur spontaneously or as a result of exposure to certain environmental factors, such as radiation or certain chemicals. Chromosomal aberrations are changes in the structure or number of chromosomes that can occur within cells.

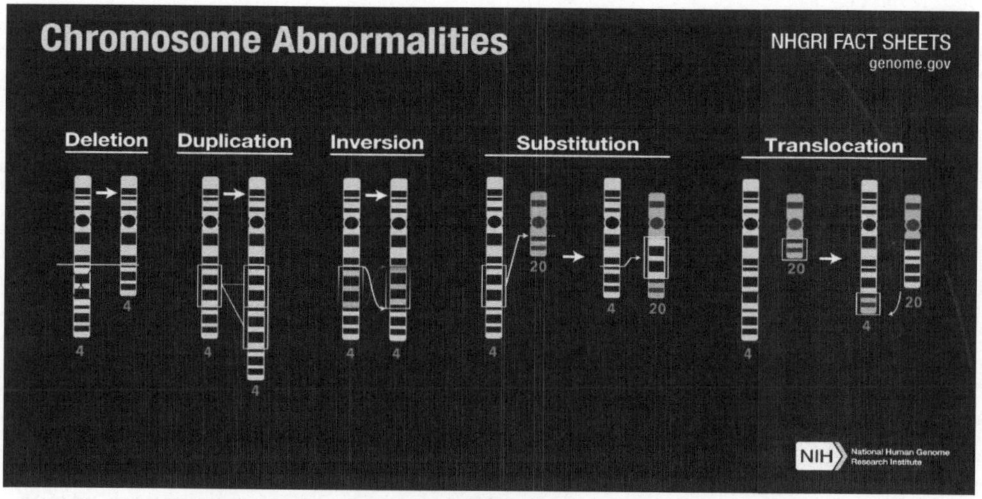

Figure 3.2: Structural Abnormalities of Chromosomes (*Source*: NIH).

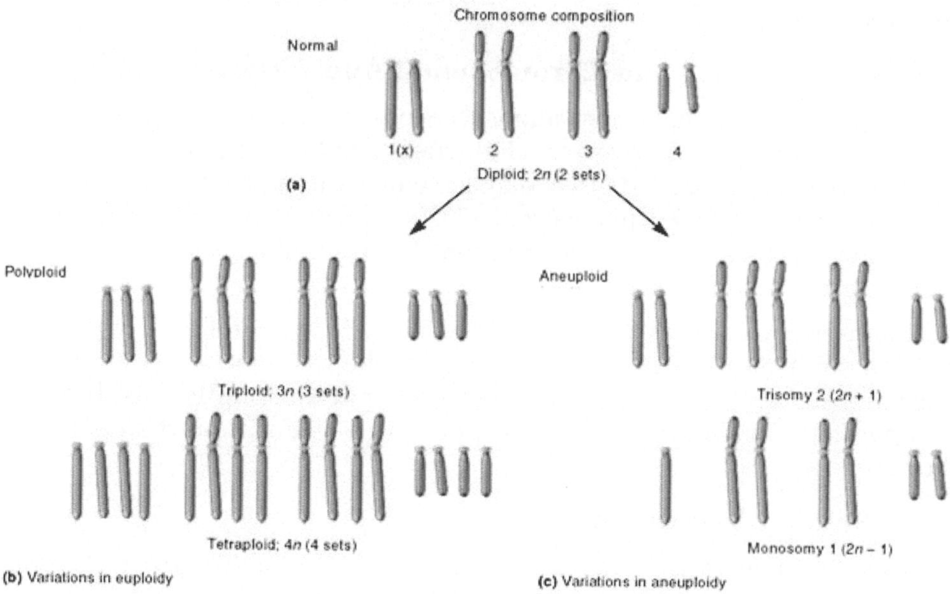

Figure 3.3: Numerical Chromosomal Aberration Types (*Source*: Plant Life).

There are several causes of chromosomal aberrations, including nondisjunction, chromosomal breakage, mitotic crossing over, meiotic non-disjunction, errors in DNA replication, and certain diseases and infections. Nondisjunction occurs when chromosomes fail to separate properly during cell division, leading to an abnormal number of chromosomes in daughter cells. Chromosomal breakage, caused by physical or chemical agents such as radiation or certain chemicals, can result in rearrangements and structural aberrations. Mitotic crossing over, which is a normal process that occurs during cell division, can result in structural aberrations if genetic material is exchanged between non-homologous chromosomes. Meiotic non-disjunction occurs when chromosomes fail to separate properly during meiosis, the process that leads to the formation of sperm and egg cells. Errors in DNA replication, such as slips or misalignments, can also lead to chromosomal aberrations. Certain diseases and infections, such as cancer or human papillomavirus (HPV), and aging can also increase the risk of chromosomal aberrations.

In conclusion, chromosomal aberrations are significant changes in the structure or number of chromosomes that can result in a range of genetic disorders. These aberrations can be caused by a number of different factors, including nondisjunction, chromosomal breakage, mitotic crossing over, meiotic non-disjunction, errors in DNA replication, and certain diseases and infections. Understanding the causes and consequences of chromosomal aberrations is important for advancing our knowledge of human genetics and improving human health.

Cosmic Radiations and Chromosomal Aberrations

Multiple chromosomal aberrations were observed in the plants grown from the seeds that were sent to space. High energy radiations like the α-rays are the cause of space mediated DNA-damage which ultimately led to aberrations (Mohanta *et al.*, 2017). Furukawa *et al.* (2020) reported unstable chromosomal aberrations such as ring, rearranged acentric, and multicentric chromosomes in space grown plants.

Space environment is also reported to have reduced significantly the mitotic division ability which lead to an increase in the frequency of chromosomal aberrations. Delone *et al.* (1965) reported change in spindle axis direction, unseparated chromatids, 3- and 4- pole mitoses and non-disjunction of chromosomes.

Copy Number Variations

Copy number variation (CNV) refers to variations in the number of copies of a specific segment of DNA among individuals. These variations can range from small differences, involving just a few base pairs, to large differences that include thousands of base pairs. CNVs can arise through duplications, deletions, or other changes in the DNA structure and can impact long regions of DNA, including those that contain genes. Microdeletions and microduplications are examples of CNVs.

Microdeletions are deletions in chromosomal regions which cannot be detected through conventional light microscopy techniques or conventional cytogenetic methods. They are typically about 1-3 Million base pair long (Mb) and might involve several contiguous genes.

Microduplications are small genetic duplications that occur in the DNA. They are typically less than 5 million base pairs in size and result from an extra copy of a segment of DNA being present in a person's genome. The impact of a microduplication depends on the location and size of the duplication, as well as the specific genes that are affected.

Chromoanagenesis

Chromoanagenesis refers to a catastrophic event that typically involves localized shattering and rearrangement of chromosomes, leading to a significant restructuring of the chromosome. Originally discovered in cancer cells, this phenomenon has since been seen in other organisms, including plants.

The broader term "Chromoanagenesis' includes catastrophic phenomena such as chromothripsis, chromoanasynthesis and chromoplexy.

Chromothripsis is caused by a singular cellular occurrence where one or more chromosome segments break into pieces and are reassembled in an arbitrary order and orientation, leading to the formation of intricate derivative chromosomes. The defining features of chromothripsis include the presence of many grouped breakpoints in the chromosomes, limited changes in the DNA copy number, and the maintenance of heterozygosity in the rearranged segments. These characteristics are seen in all chromothripsis events.

Liu *et al.*, described Chromoanasynthesis as a unique type of "one-time" disordered chromosomal rearrangement. This process involves the separation of the lagging strand and its subsequent attachment to a nearby replication fork site, resulting in multiple, complex rearrangements of varying sizes. Chromoanasynthesis can produce intricate rearranged chromosome segments with frequent changes in DNA copy number, including region-specific duplications and triplications, as well as short micro-homology segments at the breakpoint junctions.

Chromoplexy is a type of intricate DNA rearrangement seen in the genomes of cancer cells. It was initially discovered through whole genome sequencing of prostate tumors (Shen *et al.*, 2013). Chromoplexy results in the scrambling of genetic material from one or more chromosomes as multiple DNA strands are broken and connected in a new arrangement.

Techniques for Studying Chromosomal Aberrations

There are several methods used to detect chromosomal aberrations, which are changes in the number or structure of chromosomes. The choice of method depends on the type and severity of the chromosomal aberration being studied and the goals of the testing. In many cases, a combination of methods may be used to provide a more complete picture of the changes in the genome. Some

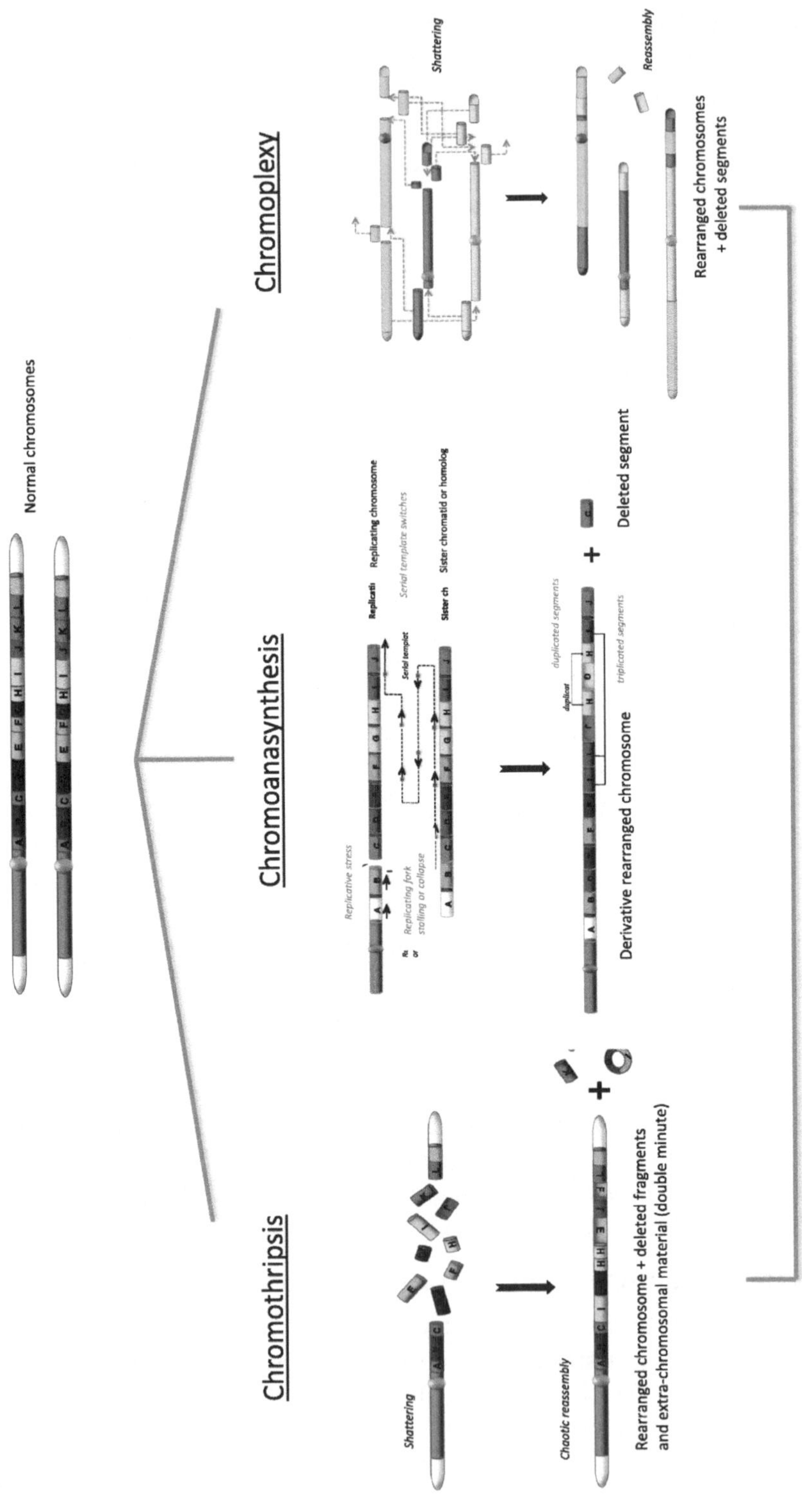

Figure 3.4: The Characteristics of Chromoanagenesis, which Encompasses the Disordered Mutation Processes of Chromothripsis, Chromoanasynthesis, and Chromoplexy. (*Source:* Franck *et al.*, 2020).

of the most commonly used techniques include karyotyping, Fluorescent in situ hybridization (FISH), array comparative genomic hybridization (aCGH), Next-generation sequencing (NGS), Multiplex ligation-dependent probe amplification (MLPA) amongst others.

Different Techniques for Detecting Chromosomal Aberrations (Source: Song *et al.*, 2017)

Technique	*Application*	*Advantage*	*Shortcoming*
Metaphase Cytogenetics (MC)	Visible chromosomal aberrations	Simple, whole chromosomal view	Low resolution, Can't detect UPD
FISH	Small and hidden chromosomal aberrations	Not rely on proliferating cells, High sensitivity	Only detect particular chromosomal aberrations
Spectral Karyotyping (SKY)	Unknown and complex chromosomal aberrations	Display better pictures of karyotypes	Can't detect structural aberration, low resolution
SNP array	Cryptic and complex chromosomal aberrations	High-resolution, can detect UPD	Can't detect balanced translocation and inversion
Array-CGH	Detect CNV and UPD	Genome-wide analysis high-resolution	Can't detect balanced rearrangements, low-level mosaicism and polyploidy
Sequencing	CNV and structural variants, unknown mutation or aberrations	Genome-wide analysis improved sensitivity monitor clonal mutations	Expensive, time-consuming, complicated bioinformatic analysis

FISH (Fluorescence *In situ* Hybridization)

It is a molecular cytogenetic technique that localizes a specific DNA sequence or an entire chromosome. It can be used to detect small alterations that range from 50 Kb to 100 Kb. The high analytical resolution and sensitivity of FISH, made it a widely used method for diagnosis of aneuploides, microdeletion/ microduplications and subtelomeric arrangements (Cui *et al.*, 2016). Procedure of FISH involves construction of a probe, which is then labelled with fluorescent dyes and then is hybridized with the subject DNA. The fluorescent signals are then detected through fluorescence microscopy. There are multiple modifications of FISH that are used in detection of chromosomal abnormalities. Some of them are ACM-FISH, armFISH, CARD-FISH, CAT-FISH, Fibre-FISH, Flow-FISH, DBD-FISH, M-FISH, PCC-FISH *etc.* to name a few (Ratan *et al.*, 2017).

Flow Cytometry

Flow cytometry is a method used to analyze and sort particles, such as cells, in a fluid stream based on their physical and fluorescent characteristics. It uses laser light to illuminate particles as they flow single-file through a detection

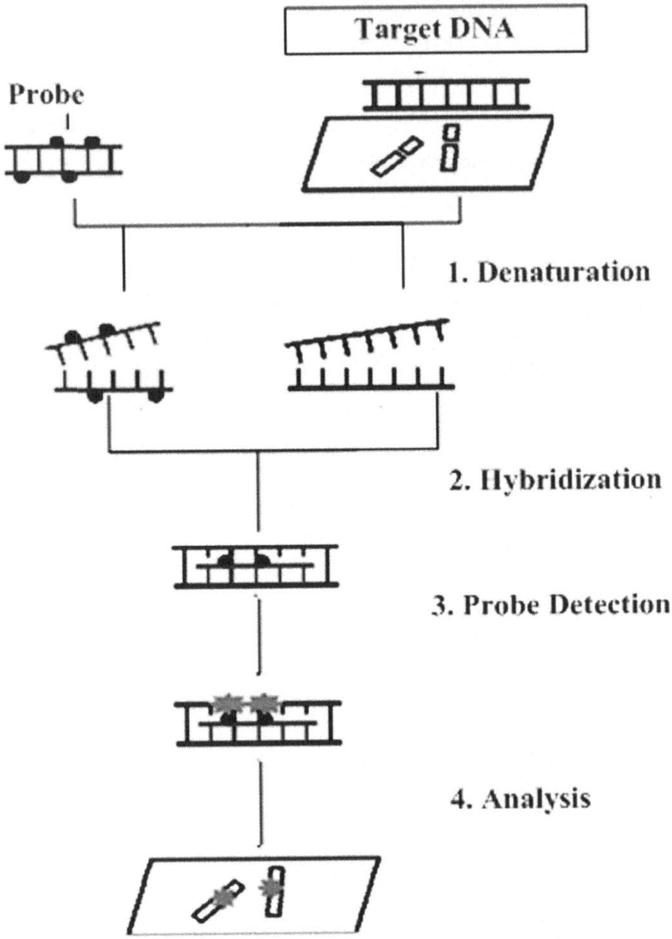

Figure 3.5: Workflow of FISH (*Source*: Ratan *et al.,* 2017).

device, and detectors measure the light scattered by the particles, as well as fluorescence emissions from any fluorescent dyes bound to the particles. The resulting data can be used to identify and quantify various cell populations, analyze cellular properties such as size, granularity, and fluorescence intensity, and sort cells into discrete populations based on these properties.

Flow cytometry is a powerful tool for cell analysis, with applications in many areas of research and clinical practice, including immunology, hematology, cancer research, and stem cell biology. It is also used in medical diagnostics for the analysis of blood and bone marrow cells, and in quality control for the production of biologic products. Flow cytometric analysis of chromosomes, also known as flow karyotyping, is a technique that allows for the detection of chromosomal abnormalities. Flow sorting also allows for the segregation of individual chromosome types in substantial amounts, facilitating gene mapping, the establishment of chromosome-specific gene libraries, and sequencing for chromosome genomics. However, it should be noted that flow karyotyping has limitations and cannot detect translocations with similar DNA content or inversions.

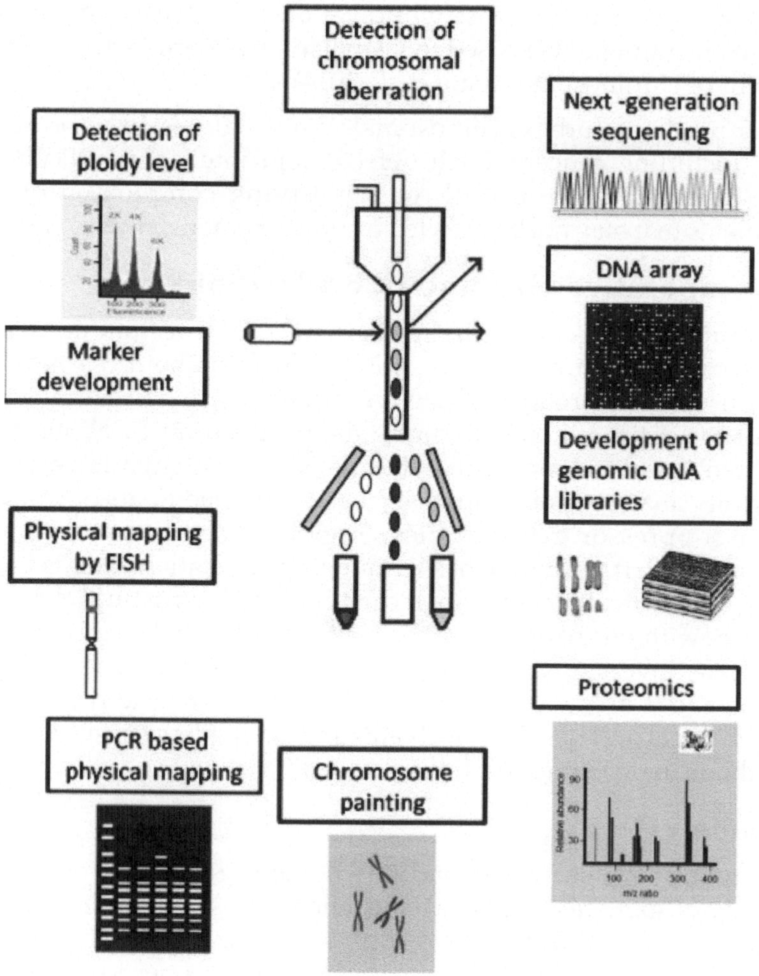

Figure 3.5: Flow Cytometry and its Applications.

Spectral Karyotyping

Spectral karyotyping (SKY) is a laboratory technique used to visualize and analyze the structure and arrangement of chromosomes. It is a type of fluorescence in situ hybridization (FISH) technique that uses multiple fluorescent dyes to label different chromosomes and create a unique "spectral" pattern for each chromosome. This allows for the identification of changes in the structure of chromosomes, such as rearrangements, deletions, duplications, and inversions. It utilizes a combination of Fourier spectroscopy, charge-coupled device (CCD) imaging, and optical microscopy to measure the emission spectra in the visible and near-infrared range simultaneously across the entire sample. This enables the use of multiple probes with overlapping spectrums. The SKY process involves three main steps: hybridization, image acquisition, and result analysis. The resulting data is used to visualize and examine the structure and arrangement of chromosomes (Guo *et al.*, 2014).

SKY is used to study chromosomal abnormalities in various genetic disorders, including cancers, birth defects, and intellectual disability. The technique is particularly useful for identifying complex chromosomal rearrangements that may not be detected by other methods, such as karyotyping.

CGH (Comparative Genome Hybridization)

The comparative genomic hybridization (CGH) technique is used for detecting changes in chromosomal copy numbers without the need for cell culturing. It offers a broad perspective of chromosomal gains and losses throughout the entire genome. By using the results from CGH, more specific molecular biological techniques like fluorescence in situ hybridization, loss of heterozygosity analysis, and sequencing can be applied to uncover oncogenes and tumor suppressor genes in these regions. The comparative genomic hybridization (CGH) technique has a number of limitations and is both time-consuming and difficult to perform. It cannot identify structural chromosomal abnormalities without changes in copy number, such as balanced chromosomal translocations, inversions, or ring chromosomes. It also does not provide information about tissue architecture. CGH detects gains and losses relative to the ploidy level and has the disadvantage of being less sensitive than PCR-based methods in detecting deletions.

Conclusion

The study of the structural and functional organization of the genome and nucleus of gametes is crucial and encompasses complex cytogenetic and molecular genetics. This field is constantly discovering new phenomena that cause chromosomal aberrations, leading to a deeper understanding of the development and evolution of plants and animals. The impact of these aberrations is particularly important in the field of human medicine, as they can lead to the development of cancers and birth defects. Chromosomal aberration assays are used to evaluate the mutagenicity of chemicals in plants using onion root tips. The analysis of numerical aberrations provides valuable information on the evolutionary history of polyploid crops. Significant advances have been made in understanding the mechanisms behind chromosomal mosaicism and complex rearrangements, as well as in their diagnosis.

REFERENCES

Baca, S. C., Prandi, D., Lawrence, M. S., Mosquera, J. M., Romanel, A., Drier, Y., Garraway, L. A. (2013). *Punctuated evolution of prostate cancer genomes.* Cell, 153(3), 666–77. doi: 10.1016/j.cell.2013.03.021

Berger, M. F., Lawrence, M. S., Demichelis, F., Drier, Y., Cibulskis, K., Sivachenko, A. Y., Garraway, L. A. (2011). *The genomic complexity of primary human prostate cancer.* Nature, 470(7333), 214–20. doi: 10.1038/nature09744

Cui, C., Shu, W., and Li, P. (2016). *Fluorescence In situ Hybridization: Cell-Based Genetic Diagnostic and Research Applications.* Frontiers. https: //www.frontiersin.org/articles/10.3389/fcell.2016.00089/full

Guo, B., Han, X., Wu, Z., Da, W., and Zhu, H. (2014). *Spectral karyotyping: An unique technique for the detection of complex genomic rearrangements in leukemia.* Translational Pediatrics, 3(2), 135–139. https: //doi.org/10.3978/j.issn.2224-4336.2014.01.02

Hastings, P. J., Ira, G., and Lupski, J. R. (2009). *A Microhomology-Mediated Break-Induced Replication Model for the Origin of Human Copy Number Variation.* PLoS Genetics, 5(1). https: //doi.org/10.1371/journal.pgen.1000327

Jones, Mathew J. K., and Jallepalli, Prasad V. (2012). *Chromothripsis: Chromosomes in Crisis.* Developmental Cell, 23(5), 908–917. https: //doi.org/10.1016/j.devcel.2012.10.010

Korbel, Jan O., and Campbell, Peter J. (2013). Criteria for Inference of Chromothripsis in Cancer Genomes. Cell, 152(6), 1226–1236. https: //doi.org/10.1016/j.cell.2013.02.023

Lee, J. A., Carvalho, C. M. B., and Lupski, J. R. (2007). *A DNA Replication Mechanism for Generating Nonrecurrent Rearrangements Associated with Genomic Disorders.* Cell, 131(7), 1235–1247. https: //doi.org/10.1016/j.cell.2007.11.037

Liu P, Erez A, Nagamani SCS, Dhar SU, Ko³odziejska KE, Dharmadhikari AV, *et al., Chromosome catastrophes involve replication mechanisms generating complex genomic rearrangements.* Cell. 2011; 146: 889–903.

Mohanta, T. K., Bashir, T., Hashem, A., Abd Allah, E. F., and Bae, H. (2017a). *Genome editing tools in plants.* Gene 8: 399. doi: 10.3390/genes8120399

National Human Genome Research Institute. (2016). Chromosome Abnormalities Fact Sheet. Genome.gov. https: //www.genome.gov/about-genomics/fact-sheets/Chromosome-Abnormalities-Fact-Sheet

Pellestor, F., Gatinois, V. *Chromoanagenesis: a piece of the macroevolution scenario.* Mol Cytogenet 13, 3 (2020). https: //doi.org/10.1186/s13039-020-0470-0

Ratan Z, Zaman S, Mehta V, *et al.* (2017) *Application of Fluorescence In Situ Hybridization (FISH) Technique for the Detection of Genetic Aberration in Medical Science.* Cureus 9(6): e1325. doi: 10.7759/cureus.1325

Shen, M. M. (2013). *Chromoplexy: a new category of complex rearrangements in the cancer genome.* Cancer Cell, 23(5), 567–9. doi: 10.1016/j.ccr.2013.04.025

Song, Q., Peng, M., Chu, Y., and Huang, S. (2017). *Techniques for detecting chromosomal aberrations in myelodysplastic syndromes.* Oncotarget, 8(37), 62716–62729. https: / /doi.org/ 10.18632/ oncotarget.17698

Storchová, Z., and Kloosterman, W. P. (2016). *The genomic characteristics and cellular origin of chromothripsis.* Current Opinion in Cell Biology, 40, 106–113. https: / / doi.org/ 10.1016/ j.ceb.2016.03.003

Chapter 4

Apomixis and its Relevance in the Modern Era

Arjun Kumar Agarwal, Sumit Kumar Mishra*

Ph.D. Research Scholar, Department of Genetics and Plant Breeding, Ranchi Agriculture College, BAU, Ranchi – 834006, Jharkhand
**e-mail: arjunagarwal016@gmail.com*

Introduction

All living organisms go through the basic life cycle process of reproduction. Plants developed adaptable tactics with noteworthy physiological versatility in order to achieve effective fertilization, proliferation, and adaptability. Plants can reproduce sexually which involves the fusion of female and male gametes with the help of self- or cross-pollination, or asexually through different techniques such as budding, fission, rhizomes, tubers, cuttings, apomixis, *etc.*

It's interesting to note that just one sort of asexual reproduction exhibits instances of seed creation like sexual reproduction. This fascinating form of asexual reproduction is called Apomixis.

Leuwen Hock initially noted apomixis in citrus seeds as early as 1719. Later, Hans Winkler noticed the apomixis event in 1908 while observing a single female *Alchornea ilicifolia* plant at England's Kew Gardens constantly producing seeds. This Australian plant was unique in that it lacked a male equivalent in England. This reminded Winkler of the unusual behavior of a sexually propagated plant: a female plant may generate seeds without assistance from a male plant. As a result, he named this phenomenon apomixis, in which the seed is created despite no male involvement. It is derived from the Greek words "apo" (which means "away from") and "mixis," which means "mixing. Apomixis is an asexual mode of reproduction that forms embryo(s) and seed(s) without the sexual fusion of male and female gametes. It describes a broad range of reproductive strategies that lead to clonal multiplication through seeds. Bypassing meiosis and fertilization, it permits plants to produce clonal seeds with the same genotype as the mother. The potential of apomixis to fix any superior genotype makes it a powerful tool for developing plants with desirable traits.

Salient Features of Apomixis

i. It is an asexual medium of seed formation.
ii. Many plant species exhibit apomixis. Compared to field crops, it occurs more frequently in fruit crops.
iii. Segregation is not permitted by apomixis. It helps to maintain the mother plant's genetic integrity.
iv. Because apomixis doesn't involve the fusion of male and female gametes, it prevents recombination. In other words, it prevents the creation of genetic variability in the population.
v. It prevents gene flow. In other words, it prohibits the combination of genes from multiple sources.
vi. Through haploid parthenogenesis, a large number of pure lines can be developed in a very short time.
vii. It prevents loss of hybrid vigor and helps in maintaining superior genotypes.
viii. It is governed by a single dominant gene in most of the cases.

Types of Apomixis

A seed is formed through a series of intricate developmental stages that are highly controlled and coordinated. The phenomenon of double fertilization involves the formation of the embryo by the union of the egg cell and the male gamete, and the development of the endosperm through the fusion of two polar nuclei with another male gamete.

Unlike sexual reproduction, apomixis produces clonal descendants that completely preserve the genetic makeup of the parent plant.

There are numerous ways to make seeds with apomictic origin since apomictic reproduction requires the formation of an embryo from an apomeiotic or somatic cell with a mother genome. Apomeiosis is a process that substitute meiosis which give rise to unreduced gametophyte.

Apomixis is classified in several ways:

Classification Based on Frequency

(i) **Obligate apomixis**: In some flowering plants, sexual reproduction is absent and seed is developed by apomixis. Such apomixis is known as obligate apomixis. Such plant species produce progeny similar to their mother plant. *e.g.* Bahiagrass (*Paspalum notatum*)

(ii) **Facultative apomixis:** In some flowering plants sexual reproduction also occurs in addition to apomixis. Such apomixis is termed as "facultative apomixis". The progenies produced are heterogeneous in nature. It has been reported in Sorghum and some other plant species. *e.g.* Sea buckthorn (*Hippophae rhamnoides*)

Classification Based on Occurrence

(i) **Recurrent apomixis**: Where apomictic lines are obtained from one generation to another, it is known as recurrent apomixis. *e.g.* Taraxacum, Poa (Blue grass), onion *etc.*

The main features of recurrent apomixis are given below:

(a) The embryo-sac arises from diploid cells of the ovule without reduction in chromosome number.

(b) The plants produced by this method have diploid (2n) chromosome number.

(c) Such plants are always fertile.

(d) This type of apomixis is represented by diploid parthenogenesis, diploid apogamy and apospory.

(ii) **Non-recurrent apomixis:** When apomictic plants are not obtained from one generation to another, it is called non-recurrent apomixis. *e.g. Solanum nigrum,* Lilium species *etc.*

The main features of non-recurrent apomixis are given below:

(a) The embryo-sac contains cells that are haploid (n).

(b) Haploids plants are produced which are sterile.

(c) Inbred lines can be developed rapidly by this method.

(d) This type of apomixis is represented by haploid parthenogenesis and haploid apogamy.

Classification Based on Origin and the Type of Cell Involved

a. Sporophytic Apomixis or Adventative Embryony

Adventive embryony, also known as nucellar embryony, sporophytic apomixis, and sporophytic budding: In this kind, the embryo grows from a cell or a group of cells that are either from the nucellus or the integuments rather than the cells of the embryo sac. The resulting embryo is diploid because the integument or nucellus cells are diploid. In contrast to the traditional embryo, such embryos do not involve the development of embryo sac. Although it can happen in a variety of plant species, adventitious embryony is most common in tropical and subtropical tree species like citrus and mango as well as Garcinia, *Euphorbia dulcis*, and other plants.

b. Gametophytic Apomixis

There are numerous ways that gametophytic apomixis occurs in angiosperms. All forms of apomixis in which the maternal seed embryo grows from the egg cell of a properly grown embryo sac without fertilization are collectively referred to as "gametophytic apomixis.". The formation of the endosperm in the megagametophyte's central cell may need fertilization.

Apospory and diplospory, are two subtypes of gametophytic apomixis.

a) Apospory

In this kind of apomixis, the first diploid cells of the ovule that are located outside the embryo sac grow into a new embryo sac without meiosis. The embryo then emerges without fertilization directly from the diploid egg cell. It is the most typical form of apomixis in higher plants. Sometimes multiple embryo sacs can also be observed.

Following are the key points concerning apospory:

(i) In 1886, De Very and Bower separately introduced the term "apospory."
(ii) The embryo is created from a diploid egg cell of an additional embryo sac, which was created from the cell that was lying outside of the typical embryo sac.
(iii) It produces embryos that are diploid.
(iv) The nucellus, integument, or archesporium cell may give rise to another embryo-sac.
(v) It is of two types, *viz.*: Generative apospory and Somatic apospory.

Generative Apospory

Generative apospory is the method of producing embryo(s) from an embryo-sac that developed from an archesporium cell. *e.g.* Parthenium.

Somatic Apospory

Somatic apospory occurs when an embryo develops from an embryo-sac that developed from a nucellus or integument cell. *e.g.* Malus, Rubus, Allium, Crepis, Poa, and Opuntia.

b) Diplospory

In a diplospory, the embryo sac develops from the megaspore mother cell either directly through mitosis or indirectly through meiosis that has been disrupted. An eight nuclei embryo sac develops in both situations. The megaspore mother cell is not subjected to meiosis in mitotic diplospory. The megaspore mother cell goes through three rounds of mitosis in order to form an eight-nucleus embryo sac. In the occasion meiotic diplospory, one of the megaspores that comprise the dyad (instead of the tetrad of megaspores in sexual life cycle) degenerates. A typical embryo sac with eight nuclei forms after three mitotic divisions, besides the fact that it is unreduced in terms of chromosome count and genetic makeup compared to the mother sporophyte.

Diplospory is further subdivided into two categories:

Parthenogenesis

In 1849, Owen introduced the concept of parthenogenesis. It describes how an embryo develops from an egg cell without fertilization. Parthenogenesis is of two types, *viz.* haploid and diploid parthenogenesis.

(i) Haploid Parthenogenesis

Haploid parthenogenesis is the process by which an embryo develops from a haploid egg cell. Inbred and pure lines may be developed quickly by doubling the number of chromosomes in haploid plants using haploid parthenogenesis. *e.g. Hordeum bulbosum, Solanum nigrum*

(ii) Diploid Parthenogenesis

Diploid parthenogenesis is the process by which an embryo develops from a diploid egg cell. There are situations when an embryo-sac forms without meiosis. All of the cells in such an embryo-sac are diploid, including the egg cell, antipodals, and synergids. It produces diploid embryos. It is particularly helpful in maintaining heterosis. *E.g.* Taraxcum

Pollen grains may be made to grow into embryos in certain flowering plants. **Androgenesis** is the process through which embryos form from pollen grains or anthers. It has been noted in various other species, including rice, datura and tobacco.

Pseudogamy: In this case for embryo is developed without fertilization of the egg cell but pollination is necessary for endosperm development by fertilizing the secondary nucleus.

Apogamy

Winkler first used the term "apogamy" in 1908. It refers to the embryo's genesis, either from synergids or from antipodal cells of the embryo-sac. There are two types: (i) The development of an embryo from haploid synergids or antipodal cells is referred to as haploid apogamy. It is useful in development of pure lines. (ii) The development of embryo from diploid synergids or antipodal cells is called diploid apogamy. Diploid apogamy is useful in conserving heterosis. It has been reported in Allium, Iris and many other plant species.

Role of Apomixis in Plant Breeding

Apomixis, or asexual reproduction via seeds, is when an embryo develops from cells of the mother's genotype rather than via genetic manipulation or syngamy (fusion of male and female gametes). It helps in maintaining the hybrid vigor in subsequent generations regardless the level of heterozygosity. It would be extremely detrimental to agricultural breeding programs to maintain heterosis through conventional methods.

Even though apomixis is present over ~400 species and ~40 families and is taxonomically dispersed over all main families of angiosperms, it is mostly absent in key crop species. In Gramineae, Compositae, Rosaceae, and Rutaceae, it is most prevalent. Among the main cereals, wheat, maize, and pearl millet have apomictic lines. Apomixis is a fascinating developmental process as well as a significant agronomic attribute with enormous promise for breeding program. Despite bypassing a few crucial processes required for the production of reproductive seeds, apomictic breeding methods differ and result in healthy seeds.

Crops that are clonally reproduced could benefit from the use of apomixis technique. Pathogens (mostly viral and endophytic) develop after several rounds of vegetative multiplication and severely restrict the output of clonal crops as well as the transfer of germplasm across nations. Tissue culture or the forced synthesis of somatic embryos have significantly enhanced clonal propagation for several vegetatively propagated plant species. These methods have allowed for the tremendous proliferation of plants, and meristem culture has even made it possible to eradicate viruses. Apomixis technique in these plants would offer the extra choice and advantage of clonal seed production, producing disease-free stock that is easier to keep and transport.

There has been little success with the introduction of apomixis-related genes from wild apomicts. As a result, synthetic apomixis has been suggested as a potential platform to incorporate apomixis in any plant. This apomixis is constructed to incorporate apo-meiosis, autonomous embryo synthesis, and autonomous endosperm development. Apomixis has exceptional capability for the seed production of crop species since it would enable the fixation of beneficial and complicated genotypes. Therefore, exploiting apomixis through innovation has long been a goal in agriculture. A thorough knowledge of the genetic causes and the molecular processes behind apomixis will be necessary in order to achieve this goal.

Detection of Apomixis

Only via a thorough screening of a sizable number of plants of a variety or a hybrid can one determine conclusively whether apomixis is present or absent. The screening requires meticulously and methodically tracing each stage of embryonic development of megaspores from the embryo-sac and embryo by microtomy of ovule.

However, it is necessary to emphasized that the only recurring apomixis *viz.*, apospory, parthenogenesis, apogamy, adventive embryony which are diploid in nature is useful for plant breeding. The simple explanation is that they are the only ones that can continue to exist for many generations since they can make healthy diploid embryos without fertilization. Nonrecurrent apomixis are intended exclusively for scholarly purposes.

Maintenance and Transfer of Apomixes

As soon as an apomictic plant is found, its inheritance must be quickly investigated by crossing a small number of flowers—perhaps only one or two—with pollen taken from typical plants and observing the segregation behavior in F2 and successive generations.

The other flowers may be widely investigated, and when they are mature, seeds can be harvested. The real character of all such plants would then become distinct only after progeny tests.

A real apomictic species will naturally give birth to mother apomictic seedlings that are easy to maintain. Regarding the transmission of apomixis, there is convincing evidence that many apomicts are hybrid in origin. However, there is absolutely no proof that apomixis may be brought on by hybridization alone.

Since there is a strong possibility that the interacting gene complexes controlling apomixis would break down, as was previously indicated, the dynamic nature of apomicts makes the situation much worse. As a result, albeit unlikely, there is still a chance that non-apomicts might develop apomixis.

Role of Apomixis in Modern Era

Seed Production in Apomictic Species

Obligate apomicts can be commercially produced from seeds using the same straightforward procedures and general guidelines as cultivars that are vegetatively multiplied. These varieties are not in risk of cross - pollination; isolation is only necessary to avoid mechanical mixing during harvesting. Because an obligatory apomict does not produce sexual progeny, all seed categories are genetically identical. If a perennial seed farm is managed free of contamination from other varieties and volunteer plants of different strains of the same species, this can be managed to keep indefinitely.

Since pollination is necessary for seed production, the application of male sterility to promote the crossing of sexual plants with pseudogamous apomicts would necessitate the restoration of fertility in the F1. Special considerations must be made when producing seeds from facultative apomicts, particularly with regard to setting standards. The breeder should first decide how much variation is suitable for a specific crop. The next step is to ascertain the type and degree of variation present in each line. Only when the line has been grown for enough generations to accumulate a record of variability is it possible to discern this data. All advanced genotypes should be assessed for variation in the region and under the conditions envisaged where hybrid seed production is envisaged since environment might alter mode of reproduction.

Synthetic Apomixis through Genome Editing to Preserve Hybrid Vigor

Clonal plants are produced by further using the evolution of MiMe genotypes (Mitosis instead of Meiosis). But there is still much to learn about the paternal ancestry and determination of the genes associated in artificial induced apomixis. Gene modifications utilizing evolving technologies in plant molecular genetics has been made possible by genome editing technologies (GETs), such as clustered regularly interspaced short palindromic repeats (CRISPR)/CRISPR-associated protein 9 (cas9), a reverse genetics tool. Several genes in significant crops have been reliably altered over the past ten years. The broad availability of GETs has made it simple to perform functional genomics investigations in crops crucial to food and nutrition security.

The reduction of hybrid vigor and other desirable qualities as a result of genetic segregation in succeeding generations is the only disadvantage of hybrid technology. The price of fresh seeds for each sowing is another significant issue for poor farmers. For the sake of preserving food security, environment protection, and livelihood, hybrid seed traits must be preserved. Researchers using cutting-edge plant breeding technology are working to address the significant difficulty of ensuring that crops transmit on hybrid traits to progeny.

Applications and Advantages of Apomixis

Mode of reproduction and genetic variability are the fundamental building blocks of contemporary agricultural output, and selection and hybridization are frequently used to create high yielding cultivars from these traits. Using traditional plant breeding strategies, crops with steadily increasing yields over time have been developed. Some of these varieties that had earlier been improved through conventional methods have more recently been given genetically modified genes that regulate agronomically advantageous aspects. Farmers' capacity to successfully use these resources is constrained in some regions of the world by ecological and/or socioeconomic constraints. While poor countries must deal with ever-rising populations and worsening food shortages, wealthier nations profit from an excess of agricultural items made achievable by the application of modern technologies. With the world population expected to settle at 8–11 billion people by the middle of the 21st century, new tactics must be investigated in order to encourage a significant rise in agricultural production in such heavily populated yet economically deprived nations (Toenniessen, 2001).

The commercial seed sector is worth of around $50 billion, according to the World Seed Industry Association. A sizeable component of this industry is made up of F1 hybrid seed, which results in plants with hybrid vigour (heterosis). The commercial seed industry accounts for around a third of the world's seed stock, public institutions account for another third, and farmer-saved seed makes up the last third. The time-consuming generation of F1 hybrid seeds is necessary to reap the rewards of hybrid vigour. For both breeders and producers, the fixing of hybrid vigour via apomixis is a desired goal. The use of apomixis to stabilise heterozygous genotypes would speed up and reduce the cost of breeding projects, with an estimated profit of more than $2.5 billion annually for the production of hybrid rice alone.

Agriculture would gain a lot from apomixis in agricultural plants, including the following:

(i) The widespread use of hybrid vigor and its fixing in agricultural plants, along with those for which hybrid breeding is still unavailable.

(ii) The fast preservation of germplasm, especially those resulting from distant crossings that are not appropriate for sexual reproduction, hence maximizing the use of genetic variety.

(iii) Seeds can be produced in vegetative propagated crops which are true to the type.

(iv) It is possible to rapidly multiply genetically identical plants without taking the risk of segregation.

(v) The maternal influence, if it exists, may be effectively exploited across generations.

(vi) Homozygous inbred lines, like those seen in maize, are capable of generating segments of diploid tissues, as well as the occasional viable gamete and seed, allowing them to grow quickly.

(vii) Apomixis offers a simple method for producing hybrid seeds. Automatic apomictic processes form the hybrid seed. No crossing is necessary. In other words, commercial seed production in apomictic genotypes is not dependent on crossing.

(viii) It aids in reducing the amount of money needed to hire laborers for crossing purposes in traditional hybrids for the generation of hybrid seed. Thus helps in getting cheaper hybrid seed.

(ix) Unlike in hybrid seed production which required proper isolation to maintain genetic purity there is no need to maintain isolation distance in apomictic lines.

(x) Apomictic genotypes are simple to handle. We only need to deal with one genotype in hybrid development utilising apomixes. The breeder must deal with three different types of material, namely A, B, and R lines, when the hybrid is created via cytoplasmic genetic male sterility.

(xi) The apomixis is regulated by genetic factors. Usually controlled by a single dominant gene. As a result, cross pollination has little effect on seed production. Therefore, it prohibits outcrossing.

(xii) As already mentioned, apomictic lines always reproduce in accordance with the traits of the mother plant. To produce hybrid seed, it is not necessary to perform crosses every year. This method of producing hybrid seeds is less expensive.

(xiii) Farmers may grow their own seed with apomictic cultivars and hybrids, which is advantageous. Farmers no longer have to buy new seed from seed firms each year.

Limitations of Apomixis

(i) Low genetic variability exists in obligatory apomictic populations. In other sense, apomictic lines are quite homogeneous. As a result, these lines have limited genetic diversity and limited adaptability.

(ii) Especially obligate types of apomixis prevent population-wide spontaneous gene flow. Recombination thus do not take place in this population.

(iii) Selection is unsuccessful in obligate apomictic lines due to the insufficient heritable variation. The genotypes of all individuals in the apomictic line are identical. It might be either homozygous or heterozygous.

(iv) The frequency of apomictic lines is relatively low. These lines are only seen in a few agricultural species. However, some cereals and fruit trees exhibit these lines more frequently.

(v) In crop improvement, obligatory apomixis is simple to employ. It is challenging to employ facultative apomixis, which may reproduce sexually and through apomixis. Identification of apomictic and sexually reproducing plants becomes challenging. In fodder crops, facultative apomixis is much more beneficial.

(vi) Compared to lines used for sexual seed generation, apomictic lines make it easier to produce seeds. Farmers don't need to buy new seed from seed firms every year because they can easily create their own. Reduced seed sales as a result will have a negative impact on the seed industry.

(vii) The seed production is very easy in apomictic lines than seed production lines than seed production by sexual means. Farmers can easily produce their own seed and they need not to purchase fresh seed every year from seed companies. This will lead to reduced seed sale resulting in adverse effect on seed Industry.

(viii) Apomictic lines can occasionally become affected by their environment and multiply sexually. It raises issues regarding the use of apomictic lines in breeding programs.

Conclusion and Future Prospects

Apomixis is a complicated, biological trait projected to have an immense effect in crop breeding programs as it shortens the time required to develop a new variety and also helps to gain more revenue. Although it is still difficult for plant biologists to comprehend the occurrence of apomixis, plant breeders are nevertheless very interested in finding ways to use it to their advantage. It is possible to further study how produced mutations and naturally occurring genes might swerve the normal sexual course towards apomixis. To fully utilise the potential of apomixis to remove breeding obstacles, a combination strategy that may address apomeiosis, parthenogenesis, and embryo development may be crucial. It has been demonstrated throughout history that the development of hybrids for various crops has aided farmers in obtaining advantages, like maximum crop yields, superior grain quality, and increased resistance to biotic and abiotic stresses. However, hybrid seed is unable to produce plants with the same qualities in subsequent generations. Farmers have therefore been left with little choice except to purchase pricey hybrid seeds each year.

Apomixis has exceptional capability for the seed production of crop species since it would enable the fixation of beneficial and complicated genotypes. Therefore, exploiting apomixis through innovation has long been a goal in agriculture. A thorough knowledge of the genetic causes and the molecular processes behind apomixis will be necessary in order to achieve this goal.

REFERENCES

Barcaccia G, Albertini E. Apomixis in plant reproduction: a novel perspective on an old dilemma. Plant reproduction. 2013 Sep; 26: 159-79.

Bicknell RA, Koltunow AM. Understanding apomixis: recent advances and remaining conundrums. The Plant Cell. 2004 Jun; 16(suppl_1): S228-45.

Carman JG. Asynchronous expression of duplicate genes in angiosperms may cause apomixis, bispory, tetraspory, and polyembryony. Biological Journal of the Linnean Society. 1997 May 1; 61(1): 51-94.

Darrigues A, Daub J, McCord K, Rasmussen C, Rouse J. Genetic analysis of apomixis.

Hanna W, Bashaw EC. Apomixis: Its identification and use in plant breeding 1. Crop Science. 1987 Nov; 27(6): 1136-9.

Hojsgaard D, Hörandl E. The rise of apomixis in natural plant populations. Frontiers in Plant Science. 2019 Apr 2; 10: 358.

http: //ecoursesonline.iasri.res.in/mod/page/view.php?id=7705

Koltunow AM, Grossniklaus U. Apomixis: a developmental perspective. Annual review of plant biology. 2003 Jun; 54(1): 547-74.

Owen R. On Parthenogenesis; or, the Successive Production of Procreating Individuals from a Single Ovum: A Discourse Introductory to the Hunterian Lectures on Generation and Development, for the Year 1849, Delivered at the Royal College of Surgeons of England. van Voorst; 1849.

Richards AJ. Apomixis in flowering plants: an overview. Philosophical Transactions of the Royal Society of London. Series B: Biological Sciences. 2003 Jun 29; 358(1434): 1085-93.

Schmidt A. Controlling apomixis: shared features and distinct characteristics of gene regulation. Genes. 2020 Mar 20; 11(3): 329.

Singh BD. Plant Breeding: Principles and Methods. Kalyani publishers.

Singh P. Essentials of Plant Breeding. Kalyani publishers

Spillane C, Curtis MD, Grossniklaus U. Apomixis technology development—virgin births in farmers' fields? Nature biotechnology. 2004 Jun 1; 22(6): 687-91.

TOENNIESSEN GH. Feeding the World in the 21st Century: Plant Breeding, Biotechnology, and the. The flowering of apomixis: from mechanisms to genetic engineering. 2001: 1.

Winkler HK. Parthenogenesis und Apogamie in Pflanzenreiche. G. Fischer; 1908.

Chapter 5

An Overview of different Breeding Methods

Kulveer Singh Dhillon[1] and Khushwinder Kaur[2]*

[1]Department of Genetics and Plant Breeding, [2]Department of Plant Pathology, CSK HPKV, Palampur, Himachal Pradesh
**e-mail: kulveersinghgpb@gmail.com*

ABSTRACT

Plant breeding is an important field of agriculture aimed at improving the genetic makeup of plants to produce desirable traits. The various breeding methods used in plant breeding include hybridization, mutation breeding, polyploidy induction, and transgenic breeding. Each of these methods has its own advantages and limitations, and the choice of method depends on the specific breeding goals and the crop species being bred. Plant breeding plays a critical role in ensuring food security and the sustainable production of plant-based products. The continued development and application of these methods will be crucial for meeting the increasing demand for food and other plant-based products in the future.

***Keywords**: Breeding, Improvement, Methods, Pollination, Selection.*

Introduction

Producing new, improved crop types for farmers to use is known as plant breeding, often known as crop genetic improvement. The new variety may have higher yield, improved grain quality, increased disease resistance, or be less prone to lodging. Ideally, it will have a novel set of qualities that are vastly superior to those of the existing types. The new variety will consist of unique gene combinations that the plant breeder has assembled from the genes present in that species' gene pool. It may contain genes that are present in other kinds of the same crop, or it may also include genes from distant plant relatives or genes that were biotechnologically inserted from unrelated organisms. The objectives of this review are to describe and critically evaluate breeding methods that have been developed for breeding improved, self pollinated and cross pollinated crops. The following are the methods of breeding self-pollinated crop plants:

Plant Introduction

Introducing a genotype or a collection of genotypes to a new location or habitat where they have never been grown before. As a result, introduction may involve new cultivars of a crop previously cultivated there, a crop species that is entirely new to that region, or a crop species that is related to a wild species. Introductions have contributed immensely to the development of crop varieties through hybridization. All the semidwarf wheat varieties are derived from crosses with Mexican semi-dwarf wheats. All but few semidwarf rice varieties possess the dwarfing gene from Dee-geo-woo-gen through either TN1 or IR 8. Thus, almost all these semi-dwarf wheat and rice varieties have been developed from crosses involving introductions. *E.g.* Introduction of lRRl rice varieties, sunflower wild species from Russia and oil-palm into Tamil Nadu. Plant introduction may be of two types:

1. Primary Introduction

When the introduced crop or variety is well suited to the new environment, it is directly grown or cultivated without any alteration in the original genotype. This is known as primary introduction. *E.g.* IR8, IR20, IR34, IR50 rice varieties, Mexican semidwarf wheat varieties: Sonora 64 and Lerma Rojo, oil palm varieties introduced from Malaysia and Mashuri rice from Malaysia.

2. Secondary Introduction

The introduced variety may be subjected to selection to isolate a superior variety or it may be used in hybridization programme to transfer some useful traits. This is known as secondary Introduction. *E.g.* In soybean EC39821 introduced from Taiwan is subjected to selection and variety Co1 was developed. In rice ASD 4 is crossed with IR 20 to get Co44 which is suited for late planting.

Plant Introduction Agencies

A centralized plant introduction agency was initiated in 1946 at IARI, New Delhi. During 1976 National Bureau of Plant Genetic Resources (NBPGR) was started. The bureau is responsible for introduction and maintenance of germplasm of agricultural and horticultural plants. Similarly, Forest Research Institute, Dehradun has a plant introduction organization, which looks after introduction, maintenance and testing of germplasm of forest trees. All the introductions in India must be routed through NBPGR, New Delhi. The bureau functions as the central agency for export and introduction of germplasm. At International level International Board of Plant Genetic Resources (IBPGR) with headquarter at Rome, Italy is responsible for plant introduction between countries.

Acclimatization

When superior cultivars from neighbouring or distant regions are introduced in a new area, they generally fail initially to produce a phenotypic expression similar to that in their place of origin. But later on they pick up and give optimal phenotypic performance, in other words they become acclimatized to the new ecological sphere. Thus, acclimatization is the ability of crop variety to become adapted to new climatic and edaphic conditions. The process of acclimatization follows an increase in the frequency of those genotypes that are better adapted to the new environment.

Selection

Isolation of desirable plant types from the population is known as selection. It is one of the two fundamental steps of any breeding programme *viz.*, creation of variation and selection. There are two agencies involved in carrying out selection: one is Nature itself (Natural selection) and the other is man (artificial selection). The effectiveness of selection primarily depends upon the degree to which phenotype reflects the genotype. Before domestication, crop species were subjected to natural selection. But in modern plant breeding methods natural selection is of little importance and artificial selection plays an important role.

Basic Principles of Selection

1. Selection operates on existing variability: The main function of the selection exercise is to discriminate between individuals. This is possible only when sufficient variation is present in the material subjected to selection pressure. Thus, selection acts on the existing variation, it cannot create new variation.
2. Selection acts only through heritable differences: Only the selected individuals are permitted to contribute to the next generation/ progenies. Therefore, should there be greater influence of non-heritable agencies on the individuals selected, the parent-progeny correlation will be greatly vitiated. Hence the variation among individuals to be selected must be genetic in nature, since it is the genetic variation that tends to close the gap between phenotype and genotype. Environmental variability cannot be of any use under selection.
3. Selection works because some individuals are favoured in reproduction at the expense of other: As a consequence of its past evolutionary history and breeding structure, a population or a crop consists of

highly genetically variable individuals with regards to such diverse phenomena as differential viability, differential maturity, differences in mating tendencies, fecundity, and duration of reproductive capacity. Hence some individuals tend to become superior to others for some or other traits desirable under domestication. These superior individuals are retained for reproduction while others discarded under selection.

Selection has two basic characteristics *viz.*,

1. Selection is effective for heritable differences only, Selection does not create any new variation and
2. It only utilizes the variation already present in a population.

The two basic requirements for selection to operate are:

1. Variation must be present in the population and
2. The variation should be heritable.

Selection intensity: The selection differential is the difference between the mean of the selected group and the mean of the population. Selection intensity is therefore the mean of the deviations from the population mean, measured in units of the phenotypic standard deviation of the population. In other words, Percentage of plants selected to be advanced to next generation from a population.

Mass Selection

Large number of plants having similar phenotype are selected and their seeds are mixed together to constitute a new variety. Thus, the population obtained from selected plants will be more uniform than the original population. However, they are genotypically different.

Mass selection involves the selecting of a group of individuals from a population on the basis of their similar phenotype in an attempt to improve the performance of the population for that character or characters. A knowledge of the genetic basis and the extent of the population variability for the character indicates to the crop breeder the potential offered for improvement through selection. In the mass selection procedure there is no progeny test (in contrast to pure line selection, and progeny selection and line breeding) but the selected population is usually evaluated against the original unselected population to gauge the effectiveness of selection.

In self-pollinated crops, mass selection has two main uses. Firstly, it is an efficient method of bringing rapid improvement to populations which have a proportion of individuals that are obviously (*i.e.* visually) unfit, such as too early or late in maturing, disease susceptible or lodging susceptible. Mass selection is used in this way to 'purify' populations for a narrower range of character value(s), a practice which is believed to confer closer-to-optimum fitness for that character, or characters, in commercial production. It is also used in pure seed production in self-pollinated crop varieties where selection is based on discarding phenotypic variants that diverge from the standard type, or range, for the variety.

In cross-pollinated crops, mass selection has contributed substantially to their improvement, particularly for characters of high heritability such as disease resistance, date of maturity and protein or oil content of grain. There are numerous examples of its successful application. Mass selection is, however, not an efficient method for improving characters of low heritability such as yield.

Merits of Mass Selection

1. Since large number of plants are selected, the variety developed has wider adaptability as compared to variety developed through pureline selection.
2. Often extensive yield trials are not necessary. This reduces the time and cost for developing new variety.
3. Mass selection retail considerable genetic variability in the new variety. Therefore, another round of mass selection after few years would be effective in improving variety further.
4. It is less demanding method so that breeders can devote more time to other breeding programmes.

Demerits of Mass Selection

1. The varieties developed through mass selection show variation and are not as uniform as pureline varieties. Therefore, such varieties are less liked than pureline varieties.
2. Improvement through mass selection is less than that could be achieved through pureline selection, because some progenies will be poorer in performance as compared to the best progeny/pureline present in new variety.
3. In the absence of progeny test it is difficult to find out whether selected plants are homozygous as there may be some degree of cross pollination in self-pollinated crops, some of the selected plants may be heterozygous.
4. Varieties developed through mass selection are more difficult to identify than pureline varieties in seed certification programmes.
5. Mass selection utilizes variability already present in a variety or population, and it cannot generate variability.

Progeny Test

Evaluation of the worth of plants on the basis of performance of their progenies is known as progeny test. This was developed by Louis de Vilmorin and so it is also known as the Vilmorin Isolation principle. Vilmorin worked on sugar beet plants. The progeny test serves two valuable functions:

1. Determines the breeding behaviour of a plant *i.e.* whether it is homozygous or heterozygous.
2. Whether the character for which the plant was selected is heritable *i.e.* due to genotype or not.

Pureline Selection

Pureline is the progeny of a single self-fertilized homozygous plant. Pureline selection has been the most commonly used method of improvement of self-pollinated crops. Almost all the present day varieties of self-pollinated crops are purelines. Pureline selection has several applications in improvement of self-pollinated crops. It is used to improve local varieties, old pureline varieties and Introduced varieties.

Pureline Theory given by Johannsen (1903)

The concept of Pureline was proposed by Johannsen on the basis of his studies with beans (*Phaseolus vulgaris*) variety called Princess. He obtained the seeds from the market and observed that the lot consisted of a mixture of larger as well as smaller size seeds. Thus, there was variation in seed size. Johannsen selected seeds of different sizes and grown them individually. Progenies of larger seeds produced larger seeds and progenies from smaller seeds produced small seeds only. This clearly showed that there is variation in seed size in the commercial lot and it has a genetic basis.

Confirmatory Evidence of Johannsen's Experiment

In the first case, he classified the seed from each pureline into 100 mg classes, and grew them separately. The mean seed weight of progenies from different seed weight class of a single pure line were comparable with each other, and with that of the parent pureline. For example, line no 13 had seed size classes of 200, 300, 400, and 500 mg. The mean seed weights of the progenies derived from these seed weight classes were 475, 450, 451 and 458 mg respectively. The second line of evidence came from selection within a pureline. From each pureline, the largest and the smallest seeds were selected to raise the next generation. In the subsequent generations, large seeds were selected in the progenies obtained from large seeds while in these from small seeds selection was done from small seeds. Six generations of selection was ineffective in increasing or decreasing the seed size. For example, after 6 generations of selection, the mean seed weight in Line No 1 was 690 and 680 mg in the progenies selected for small and large seeds respectively. Thus, selection within a pureline was ineffective. The third approach was to estimate parent offspring correlation. The value of parent offspring correlation within line no 13 was – 0.018+ 0.038, that is, zero, while it was 0.336 + 0.008 in the original seed lot of the Princess which is highly significant. The parent-offspring correlation will be zero when the variation is nonheritable, while it will be significantly greater than zero when the variation has a genetic basis, *i.e.*, is heritable.

Conclusions from the Johannsen's Experiment

A self-fertilized population consists of a mixture of several homozygous genotypes. Variation in such a population has a genetic component, and therefore selection is effective. Each individual plant progeny selected from a self-fertilized population consists of homozygous plants of identical genotype. Such a progeny is known as pureline. The variation within a pureline is purely environmental and, as a result, selection within a pureline is ineffective.

Origin of Variation in Pure Lines

Pure lines show genetic variation after some time because of the following reasons:

1. Mechanical Mixture: During cultivation, harvesting threshing and storage, other genotypes may get mixed up.
2. Natural hybridization: Through pure lines are produced in self-pollinated crops, some amount of natural cross pollination occurs in them also can be avoided by isolation and rouging.
3. Mutation: occur spontaneously in nature at random

The pure line selection method depends on the knowledge, that continued selfing of a heterozygous individual or population of a self-pollinated plant species will result in an increasing proportion of the subsequent population becoming homozygous. Eventually all individuals will be homozygous. In crop breeding the parents of a cross usually differ by a large number of gene pairs and this number will influence the number of generations of self-fertilisation necessary for the hybrid population to reach homozygosity. At this point the population would be heterogeneous and composed of a large number of different homozygous individuals. The degree of homozygosity of the hybrid population depends on the number of independent gene pairs by which the parents differ and the number of generations of selfing since hybridisation. The proportion of the population which is homozygous is given by the formula (2m-1)/2m)n where m = the number of generations of self-fertilisation; and n = the number of independent gene pairs by which the parents differ. Thus, if the parents of a cross differ by 12 independent gene pairs, the proportion of homozygosity of the hybrid population after 6 generations of self-fertilisation would be approximately 83 per cent. The pure line method is the basis of selection in breeding programs for improvement in self-pollinated crops, such as wheat and barley.

Transgressive Breeding

It aims at improving yield or its contributing character through transgressive segregation. It refers to the appearance of such plants in F_2 generation that are superior to both the parents for one or more character. It is due to accumulation favourable genes from the parents as a consequence of recombination. The parents used for crossing must combine very well and are genetically diverse. So, pedigree breeding followed by population approach are designed for production of transgressive segregants.

Pedigree Method

The pedigree method has found wide acceptance amongst breeders of self-pollinated crops. It typically involves the hybridisation of two parents, one usually a commercial variety and the other chosen because of a particular superior attribute. The breeder aims to incorporate this attribute in a genetic combination that is at least equal to the commercial variety in all characters and possessing the attribute of the donor parent. The pedigree method involves selecting single plants from segregating populations of a cross and evaluating the performance of their progeny with repeated selection within the better progeny for single plants followed again by further progeny evaluation. Single plant selection can be commenced as early as the F_2 population or it can be delayed until the F3 or F4 generation from the bulk segregation population. Single plant selection followed by progeny evaluation is followed until the F_6 or F_7 generation, by which time most of the lines are homozygous and can be judged visually for phenotypic uniformity. At the F_7 the homozygous selected lines can be placed in replicated, multi-location yield trials for further selection on the basis of yield performance. Some breeders employ F_2 single plant selection and then select single plants again at F_5 or F_6. This approach produces extremely uniform varieties but takes time especially in species with low seed multiplication rates (*e.g.* pulses).

The crop breeder usually selects in early generations, on a visual basis, for qualitative characteristics such as height, standing ability, morphology of the inflorescence, disease resistance and seed characters such as shape, colour and size. Because most of these characters are of reasonably high heritability and under simple genetic control this selection is very efficient in improving the overall agronomic suitability of the resultant population. Selection in later generations can be commenced for quantitative characters such as yield and quality. Because there are fewer selected lines and greater amounts of seed of the selections, small replicated trials are suitable for the reliable evaluation of such characters. Many studies have been conducted on the efficiency of selecting for yield on a single plant, or row, basis in early segregating generations but it is not considered to be a reliable method.

The theory behind the use of the pedigree method of selection bears upon its practicability in enabling the breeder to handle a very large number of recombinants in early segregating generations. When the breeder is using parents which differ by a large number of gene pairs, even when the character being selected is under simple genetic control, the possible number of recombinant types for the overall parental differences is very large. If the parents differ by *n* allelic pairs of genes with full dominance the kinds of possible phenotypes in the F_2 generation are 2^n. Thus, for a difference between parents of 20 allelic pairs of genes the number of possible phenotypes in the F_2 is 2^{20} (*i.e.* 1,048,576). Although the breeder can never expect to grow populations that reflect these huge numbers, experience and computer-simulation study have shown that the pedigree method offers a reasonably good chance of isolating a nearly best genotype with about 90 per cent of the desirable genes available from the two parents of a cross.

Multiple Cross

This method of breeding is similar to the pedigree method except that it involves the hybridisation of more than two parents, sometimes from 4 to 8. It has been used in attempts to combine the attributes for a particular quantitative character, such as yield or grain quality, from a number of adapted parents of divergent genetic origins. One severe limitation in the use of the method is the large number of crosses which would have to be performed especially in the eight parent cross (*i.e.* in hybridising the two double F_1's) in order to have a good chance of adequately sampling the possible range of recombinants. In other words, the F_1 parents will be segregating to produce F_2 gametes and this gamete population must be adequately sampled in the crossing to ensure that as many gene combinations as possible are represented. The subsequent selection procedure could be that shown for either the pedigree or the bulk selection method.

Bulk Method

The bulk method of breeding involves hybridising two chosen parents and growing the F_2 and subsequent generations until the F_6 generation in bulk populations as field plots in successive years. During this time selection is usually not practised but in the F_5 or F_6 a large number of single plant selections are taken and their progeny sown as individual rows in the following season. Progenies can be culled on agronomic performance, including height, standing ability, threshing characteristics and such characters as flowering time and grain protein content. The progeny rows which are selected and which appear to be homozygous, as judged on the phenotypic uniformity of the row, can be evaluated for yield in small-plot replicated trials in the following year.

The bulk method of breeding relieves the breeder of the large amount of work in sowing, selecting and harvesting single plant selections and progeny rows from F_2 to F_6 as is the case with the pedigree method. It has the advantage of enabling the breeder to handle a large number of crosses because of the relative ease in bulk sowing and harvesting the segregating populations. The comparative evaluation of bulk yields of the segregating populations of different crosses has been used to indicate those crosses wherein selection for yield might provide the highest yielding segregates.

Single Seed Decent (SSD) Method

This method is a modification of bulk method. Single seed from each F_2 plants is bulked to raise the F_3 generation. Similarly, F_3, F_4, F_5 generation when the plants are homozygous plant progenies are advanced to next generation. Selection is done mainly among the progenies and number of progenies is sufficiently reduced to permit replicated trail. Individual plants may be selected from outstanding families showing segregation. So preliminary yield trial and quality tests begin in F_7 to F_8.

Single seed descent (SSD) is used in the breeding of self-pollinated plants by advancing lines from generation to generation through the use of a single seed from the progeny of each plant (rather than single plant and progeny row selection as in the pedigree method). In most cases where SSD is employed the populations are grown as rapidly as possible in artificial conditions (glasshouse or growth room) so as to hurry through at least two generations per year and achieve near-homozygosity as quickly as possible. This is expensive and since the populations are still quite large it restricts the number of crosses that can be handled this way. Selection is rarely practised during this rush to homozygosity except for highly-heritable characters that can be cheaply and easily scored on single, crowded plants without risk of loss. Once homozygosity is approached, say F_5, the lines are bulked-up and serious selection can commence. As soon as seed supply permits, field trials for quantitative characters can begin. While there is a time saving to be made with SSD, undesirable genotypes will persist in the breeding program for a greater number of generations before being eliminated.

Back Cross Method

A Crossing between a F_1 hybrid or its segregating generation with one of its parents is known as Back cross. The hybrid and its progenies in the subsequent generations are repeatedly back crossed to one of their parents. As a result, the genotype of back cross progeny becomes increasingly similar to that parent to whom the back crosses are made. At the end of 6-8 back crosses, the progeny would be almost identical with the parent involved in back crossing.

Objective

1. To improve one or two specific defects of a high yielding variety and a well-adapted variety with desirable character.
2. The characters lacking in this variety are transferred to it from a donor parent without changing the genotype of this variety except for the genes being transformed.

Requirements of Back Cross Breeding

1. Suitable recurrent parent must be available which lacks in one or two characteristics.
2. A suitable donor parent must be available; the character must be highly intense.
3. The character(s) to be transferred must have high heritability and preferably, should be governed by one or few genes
4. A sufficient number of back crosses should be made so that genotype of recurrent parent is recovered in full. Ordinarily, 6-7 backcrosses are sufficient for the purpose.

Applications of Back Cross Breeding

1. Inter varietal transferring of simply inherited traits. Characters governed by one or two genes like disease resistance are successful.
2. Inter varietal transfer of quantitative characters and highly heritable quantitative characters like earliness, plant height, seed size and seed shape are transferred.
3. Inter specific transfer of simply inherited characters: Disease resistance is transferred from related species to cultivated species. Inter specific transfer of genes are easy when the chromosome of the two species pair regularly.
4. Transferring of cytoplasm: Wild species cytoplasmic are transferred to cultivated species transfer of male sterility. The variety or species from which the cytoplasm is to be transferred is used as the female parent. The parent to which the cytoplasm is to be transferred is used as the male parent in the original cross and back cross. After 6-8 back crosses the progeny would have the nuclear genotype of the recurrent parent and the cytoplasm from the donor parent.
5. Transgressive segregation: F_1 is back crossed to one or two times to the recurrent parent leaving much heterozygosity for transgressive segregation to appear. In the second modification two or more recurrent parent may be used in the back-cross progeny to accumulate genes from them into the back cross. Progeny of the new variety is not exactly like any one of the recurrent parents.
6. Production of isogenic lines: Isogenic lines are identical in their genotype except for one gene

Germplasm conversion: When valuable germplasm cannot be utilized in breeding programmes and may be used as recurrent parent in separate back cross programme these lines are called converted lines.

Multiline Varieties

Generally, pureline varieties are highly adapted to a limited area, but poorly adapted to wider regions. Further, their performance is not stable from year to year because of changes in weather and other environmental factors. Purelines often have only one or a few major genes for disease resistance, such as, rust resistance, which make them resistant to some races of the pathogen. New races are continuously produced in many pathogens, which may overcome the resistance present in the pureline varieties. For example, Kalyan Sona wheat (*T. aestivum*) originally resistant to brown rust (leaf rust), soon became susceptible to new races of the pathogen. To overcome these limitations, particularly the breakdown of resistance to disease, it was suggested to develop multiline varieties. Multiline varieties are mixtures of several purelines of similar height, flowering and maturity dates, seed colour and agronomic characteristics, but having different genes for disease resistance. The purelines constituting a multiline variety must be compatible, *i.e.*, they should not reduce the yielding ability of each other when grown in mixture.

In 1954, Borlaug suggested that several purelines with different resistance genes should be developed through backcross programmes using one recurrent parent. This is done by transferring disease resistance genes from several donor parents carrying different resistant genes to a single recurrent parent. Each donor parent is used in a separate backcross programme so that each line has different resistant gene or genes. Five to ten of these lines may be mixed depending upon the races of the pathogen prevalent in the area. If a line or lines become susceptible, they would be replaced by resistant lines. New lines would be developed when new sources of resistance become available. The breeder should keep several resistant lines in store for future use in the replacement of susceptible lines of multiline verities.

Population Improvement Methods for Cross Pollinated Crops

Populations of cross-pollinated crops are highly heterozygous. When inbreeding is practiced, they show severe inbreeding depression. So, to avoid inbreeding depression and its undesirable effects, the breeding methods in the crop is designed in such a way that there will be a minimum inbreeding. To isolate plants with superior genotypes. Individual plants are highly heterozygous. The progeny from such plants are highly heterogeneous and usually different from the parent plant due to segregation and recombination. This would increase the frequency of desirable genes combinations. As a result, the phenotype of the population would be favourably changed. In cross pollinated species genotype of the individual plants is of little importance but it is the frequency of desirable genes or alleles in the population as a whole that determines the value of a population. The breeding methods commonly used in cross pollinated crops are:

Simple Recurrent Selection

This method of selection involves visually selecting single plants from an open-pollinated source population and selfing them. The progenies of the selections are sown in a crossing block in the following year and allowed to freely intercross. The intercross population is sown in the following year and single plant selection and selfing carried out again. The progenies of the selfed plants are sown again in the following year in an intercross block. The second cycle of selection and sowing of progenies is called the first recurrent selection cycle. The cycle can be continued until there appears to be no further significant advance being obtained for the character or characters under selection. Because the method does not involve progeny evaluation it is suitable only for characters of high heritability.

Recurrent Selection for Combining Ability

This method of recurrent selection involves selecting single plants (called S_0 plants) from a source population, selfing them and, at the same time, crossing them to a tester. The tester is either a heterozygous population or a homozygous line, and the progeny of the test crosses to it are evaluated comparatively for the character being selected. Crosses that result in high progeny mean performance is said to have good combining ability, either for one cross (specific combining ability) or generally across all crosses involving that parent (specific combining ability). Selfed seed from S_0 plants which are good combiners is then sown in a crossing block in the field and the progenies are either intercrossed by hand or allowed to intercross naturally. The harvested seed is then sown, and a further cycle of selection and test-crossing can be carried out. Recurrent selection for combining ability has been shown to be a useful method for improving yield and adaptability of a number of cross-pollinated crops.

Reciprocal Recurrent Selection

This procedure involves the use of two genetically unrelated open-pollinated populations, A and B. Random single-plant selection (S_0 plants), and their selfing, is carried out in population A and at the same time they are crossed with a random sample of plants from population B. The same procedure is also carried out in population B. The progenies of these reciprocal crosses are evaluated in replicated trials for the character or characters being selected. Selfed seeds of S_0 plants of the A and B populations that gives superior progenies in these crosses are then sown as separate A and B populations and allowed to intercross within each population.

The bulk progenies of the intercrosses are then sown as separate A and B populations, and a further cycle of selection is carried out. It may be decided after the second cycle either to proceed to the third cycle or to use the selected populations to produce commercial hybrids. This is achieved by making crosses between the selected A and B populations.

Synthetic Varieties

A synthetic variety is produced by crossing in all combinations a number of inbreds (4-6) that combine well with each other. The inbreds are tested for their GCA. Once synthesised, a synthetic is maintained by open pollination. The lines that make up a synthetic may be usually inbred line but open pollinated variety, or other population tested for general combining ability are also be used. Synthetic varieties are common in grasses, clover, maize and sugar beets. The normal procedure is equal amounts of seeds from parental lines (Syn_0) is mixed and planted in isolation. Open pollination is allowed. The progeny obtained is Syn_1. This is distributed as synthetic variety or it may be grown in isolation for one more season and Syn_2 is distributed.

Mutation Breeding

The term mutation was coined by Hugo De Vries in 1900 for the first time and the word is derived from the latin word 'MUTARE' means to change. Mutation is the sudden heritable change other than the Mendelian segregation and gene recombination in an organism. Mutation may be the result of a change in a gene, a change in chromosome that involves several genes or a change in plasma gene. Mutations produced by changes in the base sequence of genes are known as gene or point mutations some mutations may be produced by changes in chromosome structure or even in chromosome number they are termed as chromosomal mutation.

Applications of Mutation Breeding

Mutation breeding has been used for improving both oligogenic as well as polygenic characters. Mutagenesis has been used to improve morphological and physiological characters including yielding ability. Mutations create inexhaustible variation. When no improvement is possible, this method has to be adopted. Various applications of mutation breeding are:

1. Induction of desirable mutant alleles which may not be available in the germplasm.
2. It is useful in improving specific characteristics of a well-adapted high yielding variety.
3. Mutagenesis has been successfully used to improve various quantitative characters including yield.
4. F_1 hybrids from intervarietal crosses may be treated with mutagens in order to increase genetic variability by inducing mutation and to facilitate recombination of linked genes.
5. Irradiation of interspecific (distant) hybrids has been done to produce translocations.

REFERENCES

Allard RW (1999) Principles of Plant Breeding (2nd Edition), Wiley, New York

Austin RB, Flavell RB, Henson IE and Lowe HJB (1986) Molecular Biology and Crop Improvement: A Case Study of Wheat, Oilseed Rape and Faba Beans, Cambridge University Press, Cambridge, UK

Evans D and Sharp WR (1986) Applications of somaclonal variation. Bio/ Technology 4, 528-532

Falconer DS and Mackay TFC (1996) Introduction to Quantitative Genetics (4th Edition). Longman, London

Fehr WR (1987) Principles of Cultivar Development (Volume 1, Theory and Technique. Volume 2, Crop Species), Macmillan, New York

Fehr WR and Hadley HH (eds.) (1980) Hybridization of Crop Plants. American Society of Agronomy and Crop Science Society of America, Madison, USA

Frankel OH and Hawke JG (1975) Crop Genetic Resources for Today and Tomorrow, Cambridge University Press, Cambridge, UK

Frey JK (1981) Plant Breeding II, Iowa State University Press

Hayward M, Bosemark NO and Romagosa I (eds.) (1993) Plant Breeding. Principles and Prospects. Chapman and Hall, London

Hill J, Becker HC and Tigerstedt PMA (1998) Quantitative and Ecological Aspects of Plant Breeding. Chapman and Hall, London

Hunt SP and Livesey FJ (eds.) (2000) Functional Genomics: A Practical Approach. Oxford University Press, UK

Lemerle D, Gill GS, Murphy CE, Walker SR, Cousens RD, Moktari S, Peltzer SJ, Coleman R and Luckett DJ (2001) Genetic improvement and agronomy of enhanced wheat competitiveness and weed management. Australian Journal of Agricultural Research 52, 527-548

Lui Ben Hui (1998) Statistical genomics, linkage, mapping and QTL analysis. CRC Press, Baton Rouge, USA

Mangelsdorf PC (1965) The Evolution of Maize. In J.B.Hutchinson (Ed.), Essays on Crop Plant Evolution, Cambridge University Press, Cambridge, UK

Mather K and Jinks JL (1982) Biometrical Genetics: The Study of Continuous Variation (3rd Edition). Chapman and Hall, London

McLaren JS (2000) The importance of genomics to the future of crop production. Pest Management Science 56, 573-579

Russell GE (1978) Plant Breeding for Pest and Disease Resistance. Butterworth, London

Simmonds NW and Smartt J (1999) Principles of Crop Improvement (2nd Edition). Longman, London

Singh, B. D. (2016). Plant Breeding: Principles and Methods. Kalyani publishers.

Smartt J and Simmonds NW (eds.) (1995) The Evolution of Crop Plants (2nd Edition), Longman Scientific and Technical, Harlow, London

Vavilov NI (1926) Studies on the origin of cultivated plants. Bulletin of Applied Botany Genetics and Plant Breeding 16, 1-248

Vogel KP, Pedersen and Jeffrey F. (1993). Breeding Systems for Cross-Pollinated Perennial Grasse. Agronomy and Horticulture, Faculty Publications

Chapter 6

Heterosis and its Recent Understanding

B.J. Antony[1], B.P. Maruthi Prasad[2], L. Sampath[1] and P. Dhamotharan[1]*

[1]Ph.D. Scholar, TNAU, Coimbatore – 641003, Tamil Nadu
[2]Ph.D. Scholar, UAS, Bangalore – 560065, Karnataka
**e-mail: bonipasantony@gmail.com*

Introduction

In biology, heterosis still remains a mystery as it is a result of a variety of mechanisms. The earliest scientific documentation of heterosis was made by Koelreuter (1766) on interspecific crosses in Nicotiana, Dianthus, Verbascm, Datura and other genera. Further Darwin discussed that "the first and most significant conclusion which may be drawn from the observations given in this volume is that cross-fertilization is generally beneficial and self-fertilization is injurious" in his book The Effects of Cross and Self-Fertilization in the Vegetable Kingdom. According to Beal (1876-1882), hybrids produced from maize varieties yielded 50 percent more than their respective parental variety means. Johnson explained in 1891 that "cross breeding is more effective than inbreeding, because imperfections or negative alleles intensify more in inbreeding progeny, whereas in cross breeding the different imprecation of the parents are more likely to be complemented by each other within the immediate descendants". Using inbreeding and cross breeding of maize, Shull (1908) conducted an extensive experiment on heterosis and reported the results in the article "The

Composition of a Field of Maize". Furthermore, he formulated the heterosis breeding approach for maize based on these findings.

Earlier East and Shell used the word "stimulation" and many others used the word "hybrid vigour" in discussing the phenomena of heterosis. Nevertheless, both "stimulation" and "hybrid vigour" are ambiguous and only refer to potential growth, *i.e.*, expanding size. Since the heterosis phenomenon is not always a positive one it will show a negative phenomenon also. Thus, these words are not appropriate when describing this phenomenon. Therefore, East and Hayes (1912) introduced the term "heterozygosis" to describe the loss of vigour during inbreeding and the gain of vigour after outcrossing. Further, George Harrison Shull proposed the term "heterosis" in 1914 to replace the cumbersome term heterozygosis in his invitation lecture delivered at Gottingen, Germany, as published in the paper "Duplicate genes for capsule-form in Bursa bursa-pastoris." As a result of the outbreak of World War I in 1914, the paper was published after eight years. The term "Hybrid Vigour" was suggested by Jones (1918), while studying double-cross hybrids in maize. Jones (1918) also suggested the use of double-cross hybrids for commercialization, since it produced by crossing two F_1 plants to lower seed costs. This led to the commercialization of hybrid seed production and hybrid maize farming began in the USA. B. O. Dodge (1942) also discovered "heterokaryotic vigour" in Neurospora at this period and Dr. D. F. Jones in Botanical Abstracts described heterosis as "The increase in size and vigour resulting from crossing is called heterosis".

The following hypotheses were previously proposed for the explanation of heterosis. Shull (1912) suggested that foreign sperm joining foreign cytoplasm results in an advantageous reaction in F_1, but the reaction does not repeat itself in F_2. E M East and G H Shull proposed that protoplasmic heterogeneity is associated with increased metabolic activity. D. Richey, F D Jones, and others hypothesized that heterozygotes have advantages due to linked dominant favourable factors. E M East stated that multiple alleles produced by repeated mutations at single loci produce additive effects. In 1946 Jones demonstrated the occurrence of heterosis when mutant strains were backcrossed to homozygous parental strains. Furthermore, Castle proposed sensitization of non-mutant parents by their mutated allele in 1946. A modification proposed by Dobzhansky (1950) would be to substitute the term euheterosis to address the added qualifications. The concept of heterosis was further enriched by Powers (1944) and Stern (1948) by adding the concept of negative heterosis.

Additionally, three classical theories were proposed regarding heterosis. According to the dominance hypothesis, developed by Davenport (1908) and Bruce (1910) and Jones (1917), dominant alleles are favored over recessive alleles. This concept dictates that dominant alleles present in one parent, mask the deleterious effects of recessive alleles from another parent. The overdominance hypothesis suggested by Crow (1948) and Hull (1945) suggests that heterozygote (F_1) is superior due to the interaction of different alleles compared to both the homozygotes. There is an epistasis hypothesis (Powers 1944; Williams 1959) where heterosis occurs when alleles interact at different loci.

Despite the effort and models made by several researchers in the past decades to decipher the mechanisms underlined in heterosis, its actual cause is poorly understood. The three classical models are most agreed models among several hypothesis proposed earlier to explain genetic basis of heterosis. At present, efforts are being made to unravel the molecular basis of heterosis in order to utilize it more efficiently in crop improvement. While breeders are working on improving inbreds and defining the best parental combinations, many questions remain concerning the genetic and molecular causes of heterosis. A revival of heterosis research is currently taking place, and new genomic technologies, such as microarrays, are being exploited to understand the molecular basis involved in heterosis.

Conventional Genetic Models of Heterosis

There has been a lot of discussion regarding the genetic mechanism behind heterosis since the turn of the century. In some instances, genetically diverse two homozygous lines that produce viable F1 may produce heterosis. Heterosis has been reported in many species due to increased fitness in heterozygote conditions, however the genetic basis of this phenomenon is not well understood. As a basis for explaining heterosis in classical theories, heterozygosity and homozygosity at specific loci are used. Genetic models such as dominance, overdominance, and epistasis are widely accepted classical genetic hypothesis for explaining heterosis (Lewontin, 1964). The dominant and over dominant hypotheses are based on single locus models in which a dominant allele with different levels of dominance suppresses the effect of a deleterious recessive allele leading to heterosis. It is understood that the epistatic model is a multi-locus model in which nonallelic interaction of genes is explaining the phenomenon of heterosis. A better understanding of heterosis can be gained by using these three models. There is likely no single model that can fully explain hybrid vigour or heterosis. These theories do not necessarily conflict with each other, and it is possible that in various species or in different combinations of crosses, different mechanisms may explain heterosis.

1. Dominance Model

According to dominant complementation theory, the heterosis that results from crossing two genetically diverse parents is the result of the favourable dominant allele complementing the deleterious recessive allele. For example, F_1 (A1A2B1B2) is derived from crossing P_1 (A1A1B2B2) and P2 (A2A2B1B1) where A1 and B1 are dominant alleles, A2 and B2 are recessive alleles. In the absence of interaction between the A and B loci, the heterozygote F_1 should have better performance due to the compensation of A2 and B2 by their respective dominant alleles in F_1. Davenport, 1908, reported that the dominant allele over a deleterious recessive allele increased the performance of hybrids. Keeble and Pellow further verified the hypothesis by stating that favourable dominant alleles from both parents accumulate together in F_1 and suppress the expression of deleterious recessive alleles. Using a mathematical model, Bruce (1910) explained the dominance hypothesis of population-based F_1. Mathematically, the frequency of deleterious recessive alleles in homozygote state is always lower in F_1 than in the parental populations. Using a general formula, Falconer (1989) summarized the relationship between dominance and heterosis.

$$H_{F1} = dy2$$

Where HF1 is the deviation of hybrid from midpart value, "d" and "y" are degree of dominance and genetic distance respectively.

It demonstrates that when complete dominance "d=1" and opposing type of gene distribution "y=1" in parental population was observed, maximum heterosis can be achieved.

Pseudo-overdominance hypothesis is a variant of dominance model, which explains that heterosis occurs because of two linked genes which are in repulsion phase. The complementation between closely linked dominant favourable allele and deleterious recessive allele in repulsion phase leads to heterosis. Genetic recombination will lead to the dissociation of alleles that were initially in the repulsion phase, so heterosis associated with pseudo-overdominance will not appear in the self-pollinated offspring (F_2).

There were many objections raised against the dominance complementation hypothesis. It should be possible as per theory to develop a line that has dominant alleles at all loci, which would have the same performance as an F_1 hybrid. However, it could not be developed to prove this hypothesis. This objection was explained by Jones (1917) dominant of linked gene hypothesis, which states that dominant alleles may be linked with lethal recessive alleles, and thus the probability of accumulating all the desirable dominant genes in one line is low. Hybrids perform much better than parental sums in reality, this is intended as an objection the model as it implies the hybrid should be equal to the sum of its parents' performances. Another objection is the normal distribution in F_2, which is expected to be skewed according to the dominance model. Nevertheless, a G x E interaction and the presence of more genes may contribute to the normal distribution in F_2.

2. Overdominance Model

The overdominance model states that the allelic interaction in one locus results synergistic effect in fitness and vigour, thus the interaction between different alleles in the heterozygous F1 surpasses the effect of homozygous genotypes in the inbred parents.

This hypothesis was proposed independently by Shull and East in 1908 and later this model was supported by Hull (1945). Although Shull and East, in

1908, proposed that heterozygotes are superior to homozygotes at a locus, the term overdominance was first used by Hull (1945). According to this theory, heterozygosis and heterosis are directly proportional.There are two possible reasons why heterozygotes are superior to homozygotes:

1. Production of superior hybrid substances in the heterozygote is completely different from either of the homozygous products or due to
2. The heterozygote has a higher buffering capacity due to the cumulative action of divergent alleles.

East further elaborated this theory by introducing multiple alleles at a locus which have additive effect. These additive effects are more prominent in heterozygotes with divergent alleles. Therefore, the heterosis of combinations with divergent alleles is thus higher than that of combinations with less divergent alleles. In this example, the combination of A1, A4 will demonstrate higher heterosis than combinations such as A1, A2, A3 and A4. Because A1 and A4 are more diverse than any other combination (A1 > A2 > A3 > A4).

The identification of cases of "single locus over-dominance" such as the *SINGLE FLOWER TRUSS* (SFT) gene in tomato is one of the most exciting developments in understanding over-dominant heterosis. Earlier the lack of single locus overdominance was considered as a potential objection of overdominance hypothesis. However, recent findings clearly explain single locus overdominance. The single locus overdominance explains why heterozygote "A1A2" is superior to both homozygous lines "A1A1" and "A2A2", due to interactions between A1 and A2 alleles in the heterozygote (Figure 1). An SFT gene of tomato related to FLOWERING TIME (FT) is best example, which can increase tomato yield by up to 60 per cent in a heterozygous state (Krieger *et al.*, 2010). The model plant *Arabidopsis thaliana* has also been observed to exhibit a single locus heterosis (Meyer *et al.*, 2010; Smith *et al.*, 2011).

3. Epistatic Model

The interaction between the alleles of different locus is called epistasis. Gowen (1952) suggested that such non allelic interaction may involve in heterosis. Powell (1944) and Williams (1959) proposed that the favourable epistatic interaction between non-allelic genes involved in enhancing heterosis. For example, in Figure 6.1, A1 allele as epistatic effect on unlinked B1 allele, thus when they both combined in F_1 results heterosis. There are three types of non-allelic interaction were additive x additive, dominance x dominance and additive x dominance. Majority of the traits are multigenic and several genetically controlled inter linking pathways are involved in complex traits expression. Thus, the epistasis may contribute greatly to the heterotic expression in F_1. Sometimes the epistatic QTL expressing heterosis may mimic the overdominance QTL, particularly an allele that is epistatic to another locus in trans will mimic an over-dominant heterotic QTL.

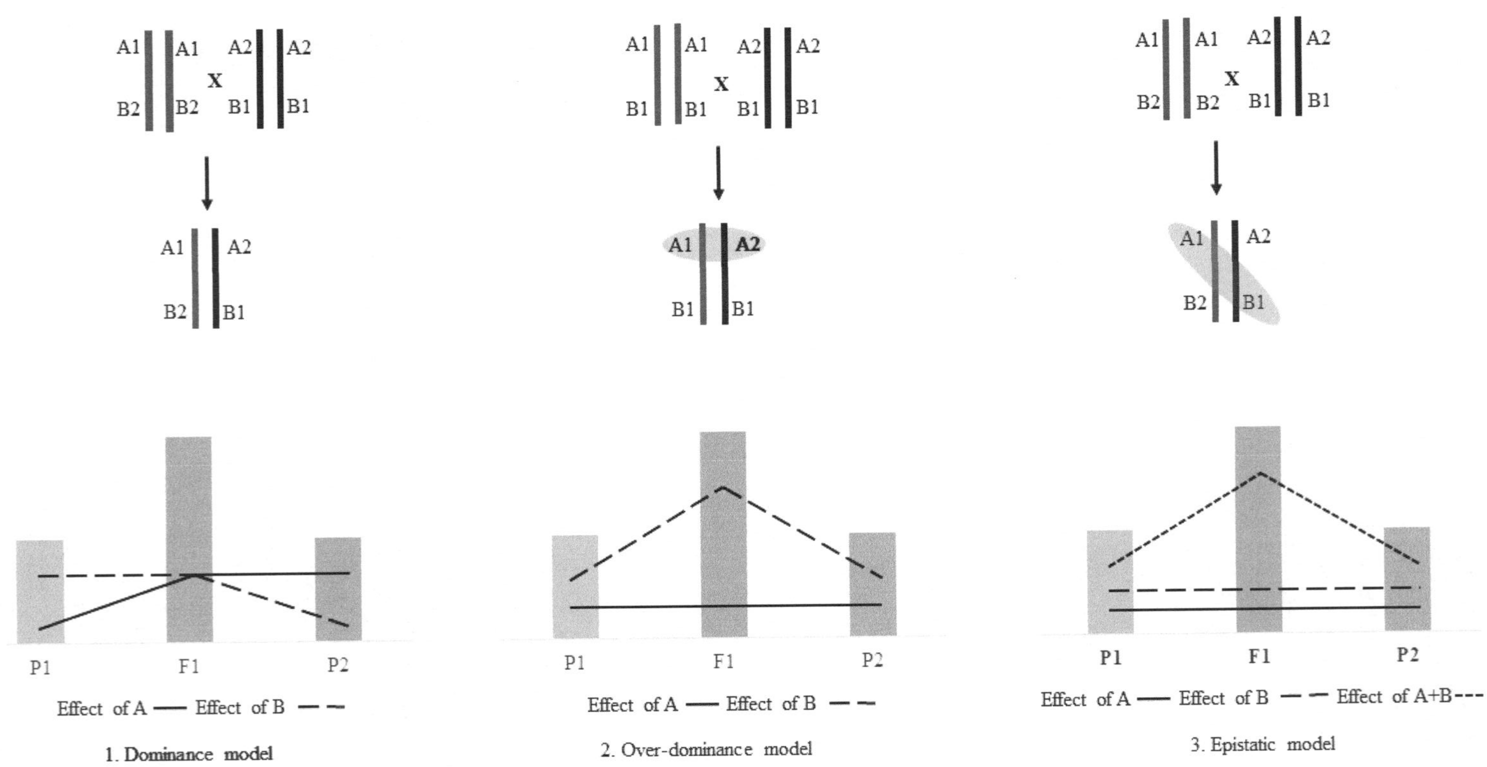

Figure 6.1

1. Dominance model - complementation of A1 and B1 allele to their recessives A2 and B2 in F_1. 2. Over-dominance model – a single gene overdominance in F_1 (A1A2) results heterosis. 3. Epistatic model – the unlinked loci A and B were interacting, where A1 and B1 interaction results in heterosis.

Genetic Distance and Heterosis

The genetic differences between parents are shown to positively correlate with heterosis in several crop species. The level of divergence between the parental lines involved in the cross, determines the magnitude of heterosis. For example, the interspecific hybrids in cotton were showed higher magnitude of heterosis compared to the intraspecific hybrids. Whereas, in rice the higher heterosis correlated with hg genetic divergence in indica x indica cross, not in indica x japonica cross. In maize studies showed the parental diversity highly correlate with heterosis. The genome wide studies in maize showed at least 10 per cent of genes present in one inbred parent is missing in another inbred parent, suggesting that the complementation results in superiority of hybrids.

Earlier the phenotypic data was used to detect the divergence, where as now the molecular markers used to detect genetic distance. Several studies in various crops suggests that the genetic distance based on molecular marker positively correlated with heterosis. However, fewer studies reported genetic distance based on random amplified polymorphic DNA (RAPD), SRS and cleaved amplified polymorphic sequence (CAPS) not correlated with heterosis in F_1 hybrids. Thus, the genetic distance based molecular markers alone cannot be used to predict heterosis. However, it is expected that molecular markers will be more useful to predict heterosis with the development of high-throughput sequencing technology. Furthermore, it appears that molecular markers' ability to predict crossing performance is strongly influenced by plant materials, phenotype, prediction methods, and planting environment. Genetic distance cannot accurately predict hybrid performance unless the DNA markers used in the analysis are associated with the genes affecting relative traits.

Epigenetics and Heterosis

The three classical theories can provide an understanding of the genetic basis of heterosis. However, these hypotheses fail to provide a complete insight about heterosis mechanism. The superiority of F_1 hybrids is the result of all genetic interactions. As genetics and genomics research advances, we gain a more comprehensive understanding of the genetic causes of heterosis. Molecular mechanisms of gene regulations could alter the fitness and growth hybrids. Epigenetic modifications are key aspects in regulating gene expression pattern, they can alter complex metabolic pathways in hybrids and causes hybrid vigour. DNA methylation, histone modification and small RNA (sRNA) are the major epigenetic factors involved in gene expression regulation. These epigenetic mechanisms do not alter the DNA sequence, but cause differential gene expression of the same individual. Consequently, this alters the phenotypic behaviour of a same individual in different condition.

1. DNA Methylation and Heterosis

DNA methylation refers to addition methyl group to the cytosine residue at its 5′ position through methyltransferase enzyme. This phenomenon is observed in majority of the plant species and high frequently at CG site of plant genome. Differential methylation patterns were observed between the F_1 hybrid and its corresponding parent. In *Arabidopsis thaliana* plant the parents showed 18-26 per cent methylation while, its reciprocal hybrid exhibited greater methylation percentage of 36-37 per cent. The reduced activity of *METHYLTRANSFERASE-1* in *Arabidopsis* correlated to reduction in the heterosis of hybrids. This suggests that the reduction in methylation level is associated with decline in heterosis. The diversity between the parents causes changes in methylation patterns of hybrids. It is noticed that more divergence between the parents causes increase the degree of methylation in F_1 hybrids. Sinha *et al.* (2020) analysed pigeon pea hybrids and their parents for differential methylation patterns. The plant immunity and growth-related genes expressed higher methylation in two hybrids ICPH 2671 and ICPH 2740 compared to the parents. In similar study, Xiong *et al.* (1999) reported that higher methylation percentage 18 per cent in F_1 hybrids of rice compare to its parent 16.3 per cent. These methylation variations alter the transcriptome profile of the hybrids and increases the growth and fitness of hybrid plants.

2. Histone Modification and Heterosis

Another notable epigenetic modification is histone modification includes acetylation and methylation of histone proteins. The covalent modification in histone octamer causes nucleosome rearrangement and altered gene expression. Histone lysine methyl transferases and demethylases modifies histone protein by adding or removing methyl group to lysine residues. Whereas, the acetylation of histone is carried out by acetyltransferase and deacetylase enzymes. In diploid and allotetraploid F_1 hybrids of *Arabidopsis*, histone modification in circadian clock related gene shows increase in heterosis. The circadian clock is a biochemical mechanism which regulates gene expression and protein production by matching with external factors such as day light. Histone modification in the promotor region of two key genes *CIRCADIAN CLOCK ASSOCIATED 1* (CCA1) and LHY involved in the regulation of circadian cycle process was studied in F_1 hybrid. These two genes downregulate several downstream genes involved in photosynthesis, bio mass accumulation. Changes in histone H3-Lys 9 acetylation and histone H3-Lys 4 dimethylation patterns in promotor regions of CCA1 and LHY genes reduces their activity in F_1 hybrids results in activation of downstream genes which increases biomass accumulation and photosynthetic efficiency. Histone modification observed in allele specific or tissue specific manner in rice (Guo *et al.*, 2015). The global allele specific expression assay in rice revealed allele specific expression is primarily regulated by allele-specific histone modifications–histone H3 lysine 36 rather than histone H3 lysine 27. The histone H3 lysine 27 also involved in parent specific expression differences in *Arabidopsis*.

3. Small Non-coding RNA (sRNA) and Heterosis

The small non-coding RNA called sRNA with 20-27 nucleotides involved transcriptome regulation also a part of epigenetic regulation mechanisms. These sRNA likely two types micro-RNA (miRNA) and small interfering RNA (siRNA) are involved in altered gene expression *via* post transcriptional gene silencing. The double strand miRNA was cleaved by *DICER LIKE 1* (DCL 1) and produce mature miRNA, which will bind to RNA Induced Silencing Complex (RISC) complex, accompanied by the *ARGONAUTE 1* (AGO1) endonuclease. The loaded complex further search for complementary mRNA and cleaves it. This way the miRNA blocks translation of specific mRNA and downregulates the expression of those genes. In a similar manner the small interfering RNA also prevents translation, where the double strand siRNA originate from transposons, repeats or RNA viruses. The differences of siRNA and miRNA levels in parents and F_1 hybrids regulated differential expression of genes and over all stability of genome. Number of studies reported siRNA and miRNA levels are substantially varying among parents and F_1s in *Arabidopsis*, rice, maize, and wheat (Wu *et al.*, 2021).

Finally, epigenetic modifications, including DNA methylation, small RNA, and histone modification, contribute greatly to heterosis. In general, hybrids with heterosis have higher methylation levels than their parents, but not all high levels of methylation lead to heterosis. Heterosis is strongly related to changes in methylation pattern and level. Hybrids may have different levels of small RNA expression and their regulatory effects on target genes, resulting in varying biomass and grain yields. Heterosis was also found to be associated with certain histones, including H3K9ac, H3K4me2, and H3K4me3. Moreover, different types of histone modification can affect genes involved in photosynthesis, circadian rhythms, and metabolism, resulting in the heterosis of hybrids. As a result of the large number of histones and the different modifications they undergo, further research is needed to determine the regulatory mechanism between histones and heterosis.

Molecular Basis of Heterosis

The superiority of hybrids over parents is studied majorly for quantitative traits like crop yield in several crop. Thus, several researchers used Quantitative Trait Loci (QTL) mapping to identify heterosis related regions in genome to understand the molecular aspects of heterosis. The QTL mapping of heterotic locus were conducted in several crop species including rice, maize, *Arabidopsis*, cotton, and tomato. The QTL are tended to express additive, epistatic, dominance and overdominance gene effect in different crops for various traits. Despite the presence of various gene action QTLs, in general, the overdominance QTL were observed to contribute significantly to heterosis. These over dominant effects were utilized in crop improvement of tomato and maize. However, there were no single QTL associated with heterosis was cloned, because it is important to know the genic level effect of over dominant loci to use them in breeding programmes. The modern tools such as Next Generation Sequencing (NGS) platforms and high through put markers like Single Nucleotide Polymorphism (SNP) help to scan entire genome of crop species to decipher genetic regions responsible for heterosis and allow researchers to clone the heterotic QTLs.

Heterotic potential of hybrids results from differential expression of genes in hybrids and parental lines. The molecular techniques allow us to study the whole genome expression profile at transcript or protein level. Transcriptome analysis refers to study of the entire transcript profile of an individual at given time. The modern techniques such as DNA Micro Array, RNA- Sequence techniques used to transcriptome analysis and gene expression studies in hybrids and parents. Various studies showed different transcript expression in parents and hybrids in crops like rice, maize, cotton, and *Arabidopsis*. These studies suggested that the favourable genes were expressed in high frequency in hybrids compared to the parental lines. The additive and non-additive type of gene expression in hybrids and parents are critical aspects responsible differential expression of genes. The proteome differences are playing a vital role in heterosis development. The different isoforms of the proteins called as isozymes could explain heterosis in some situation. In heterozygous condition the different isoforms encoded by same locus may cause hybrid vigour by altering different metabolic pathways. All activities of the entire genome result in heterosis, and transcriptome and proteome changes at the whole-genome level alter several metabolic and biochemical pathways and result in heterosis. Since, heterosis is not explained alone by either transcriptome or proteome, multi-omics approaches need to be followed to understand it.

Conclusion

Heterosis is a complex phenomenon, which is explained by several genetic aspects. There are several factors that play a role in heterosis formation, including expression patterns, environment, and physiologically-related genes. However, several questions regarding the molecular basis of heterosis remain unanswered. Novel tools and techniques are available to decipher more molecular aspects of heterosis, and understanding more molecular aspects helps to increase crop yield even more. The parental combination and combining ability of parental lines are other important and critical aspects in practical plant breeding to utilize heterosis.

REFERENCES

Chahal, G. S., and Gosal, S. S. (2002). Principles and procedures of plant breeding: biotechnological and conventional approaches. Alpha Science Int'l Ltd.

Falconer, D. S. (1989). Introduction to quantitative genetics 3rd (ed.) Longman Scientifical Techniqual.

Guo, Z., Song, G., Liu, Z., Qu, X., Chen, R., Jiang, D.,. and Yang, D. (2015). Global epigenomic analysis indicates that epialleles contribute to Allele-specific expression via Allele-specific histone modifications in hybrid rice. BMC genomics, 16, 1-9.

Hull, F. H. (1945). Recurrent selection for specific combining ability in corn 1. Agronomy Journal, 37(2), 134-145.

Lewontin, R.C., 1964. The interaction of selection and linkage. I. General considerations; heterotic models. Genetics, 49(1), p.49.

Shull, G. H. (1908). The composition of a field of maize. Am. Breeders Assoc Rep 4: 296–301.

Sinha, P., Singh, V.K., Saxena, R.K., Kale, S.M., Li, Y., Garg, V., Meifang, T., *et al.* (2020). Genome wide analysis of epigenetic and transcriptional changes associated with heterosis in pigeonpea. *Plant biotechnology journal, 18*(8), pp.1697-1710.

Wu, X., Liu, Y., Zhang, Y., and Gu, R. (2021). Advances in Research on the Mechanism of Heterosis in Plants. *Frontiers in Plant Science, 12,* 745726.

Xiong, L. Z., Xu, C. G., Saghai Maroof, M. A., and Zhang, Q. (1999). Patterns of cytosine methylation in an elite rice hybrid and its parental lines, detected by a methylation-sensitive amplification polymorphism technique. Molecular and General Genetics MGG, 261(3), 439-446.

Chapter 7

Crop-Specific Breeding Methods (Field Crops)

L. Sampath[1]*, T. Kalaimagal[2], B.J. Antony[1] and M. Himakara Datta[1]

[1]Ph.D. Scholar (GPB), Centre for Plant Breeding and Genetics, Tamil Nadu Agricultural University, Tami Nadu
[2]Professor (PBG) and Head, Department of Oilseeds, CPBG, TNAU, Coimbatore – 641003 , Tami Nadu
**e-mail: sampathabhiram3@gmail.com*

Introduction

By increasing production per unit area, primary production must practically double, at least to 80 per cent, in order to ensure global food security by 2050. Plant breeding is really difficult. Convincing the public of this issue is vital since they are already dealing with issues like climate change, a lack of suitable arable land, pressures on land for biomass production, a lack of water and phosphorus, as well as rising animal agriculture. The extremely small number of significant crops and the slow rates of breeding advancement in self-pollinating cereals raise issues in terms of breeding. When so many environmental issues are being highlighted, society and government might easily become side-tracked from the urgent need to invest in fundamental breeding research and breeding applications. It's crucial to handle these issues holistically(Peter stamp and Richard visser 2012).Here, the emphasis is placed on both the challenges to be addressed and the opportunity to achieve quality germplasm by 2050, therefore ensuring global food security. This may be done by utilising modern technology and genetics in addition to the ongoing improvement of more conventional breeding techniques. Modern agriculture generates enormous harvests every year, with just a few continents experiencing failures—almost a miracle that society takes for granted.

Plant breeding has benefited humanity from the dawn of time. Numerous variations have emerged in various agricultural plants using the principles of classical genetics. Strengthening the food front is necessary to counteract the worldwide panic caused by population growth, which presents a significant challenge to experts who study agriculture. The instruments of breeders have improved thanks to developments in molecular biology, which have also improved their confidence in their ability to help mankind. Genetically modified crop plants have previously been tested in the field as a result of the application of biotechnology to field crops. Before the end of the 20th century, it is anticipated that varieties of genetically modified rice, maize, soybean, cotton, oilseed rape, sugar beet, and alfalfa would be commercially available. It is possible to anticipate that genes from other creatures may improve crop performance, particularly in terms of their tolerance to biotic and abiotic challenges. Additionally, it is anticipated that agricultural plants will be cultivated in order to collect valuable substances like medicines made by genes inserted into them through genetic engineering. It should be noted that transgenic *Brassica napus* is already used in Europe to manufacture the anti-thrombin protein hirudin. Plant breeding seeks to enhance a plant's traits to make it more valuable commercially and agronomically. Depending on the crop under consideration, the precise aims may differ significantly (Cattivelli *et al.*, 2008). Plant breeding is the science, art, and technology of modifying a plant's genetic composition to maximise its potential for economic benefit to humankind or Plant breeding, commonly referred to as crop improvement science, focuses on the genetic improvement of plants (Jauhar, 2006).

Future developments at the global level are showing a greater interest in food security, which will have a significant impact on cereal research (Philippa 2020). For thousands of years, cereal grains have formed the mainstay of the human diet and have greatly influenced human culture. The everyday life of billions of people depends on the availability of major staples like rice, wheat, and maize, as well as, to a lesser extent, sorghum and millet (Muhammad 2013). Crops that produce oil seeds are crucial to the agricultural economy. They have a significant role in industrial uses and the human nutrition, but they are also becoming more significant as a fossil fuel substitute for supplying energy. In terms of acreage and production, oilseed crops are second only to food grains in importance to the agricultural economy. Due to rising per capita oil consumption in our diets and its usage in biofuels, the demand for vegetable oils is rising globally. Vegetable oil demand is anticipated to increase by more than twofold by the year 2050 (Savadi *et al.*, 2017).There has been little study on enhancing the nutritional qualities of pulse crops, with the focus instead being on agronomic features like yield. Compared to vegetables and cereal crops, pulses are a rich source of numerous micronutrients, particularly iron and several B vitamins (Robinson *et al.*, 2019). Despite their great nutritional value in terms of protein, resistant starch, and micronutrients, as well as their environmental sustainability as crops, pulse crops are now underutilised and only play a minor role in human diets. Because of their specialised character, research on these crops has lagged behind that of cereals (Domoney *et al.*, 2013). For billions of people worldwide, pulses represent a significant source of nourishment (Singh, 2017). Fiber crops are plants that are deliberately grown for the production of fiber for textile (clothes), cordage (*e.g.*, ropes), and filling (*e.g.*, stuffing upholstery and mattresses). cotton,jute, kenaf, industrial hemp, sun hemp, and flax are among the well-known fiber crops (Mafongoya and Sileshi 2020). A nation's economy heavily depends on its fibre crops. The principal fiber-producing crops include cotton, jute, sunn hemp, *etc.* Enhancing our ability to broadly interview nucleic-acid-dependent knowledge in the cell is critical for genomics-assisted breeding. In this sense, molecular markers have proved to be extremely useful genetic instruments for generating genetic, physical, and trait maps in a variety of fibre crop organisms (Yadav, 2022).

Breeding for Cereals Crops

The word "cereal" is derived from the Latin word "cerealis," which means "grain" and refers to a specific kind of fruit called a caryopsis, which is made up of the endosperm, germ, and bran. The annual common grass members of the grass family are the cereals (a monocot family Poaceae, also known as Gramineae**)**. Consumption of cereal grains accounts for more than 50 per cent of the global daily calorie intake. Milling removes the bran (pericarp) and germ from the majority of grains used for human consumption (Muhammad 2013).

Recently, a paradigm in cereal crop breeding has emerged that aims to combine high agronomic and biochemical qualities in a single cultivar (Philippa 2020). Studies focused at locating the desired allelic variations and quantitative trait loci (QTLs) that impact the primary breeding traits are now important. Genotyping by sequencing, commonly known as GBS, is a technique used in the area of genetic sequencing to find single-nucleotide polymorphisms (SNP) in order to carry out genotyping investigations, such as genome-wide association studies (GWAS). To hasten the breeding of desired genotypes, traditional plant breeding methods are complemented with SNP markers used for genome-wide associative studies (GWASs) and genetic editing, as well as high-throughput phenotyping techniques (Philippa 2020; Loskutov 2021).

The new breeding strategies require a supply of genes to increase the composition of the advantageous components in order to enrich cereal grains with functional components. Cereals crops are Rice (*Oryza sativa*), 2n = 24, Family: Poaceae; Wheat (*Triticum aestivum*) 2n = 6x = 42, Family: Poaceae; Maize

(*Zea mays*), 2n = 20, Family: Poaceae; Sorghum (*Sorghum bicolor*), 2n = 2x = 20, Family: Poaceae; Pearl Millet (*Penisetum americanum*), 2n=14, Family: Poaceae; Finger Millet (*Elusine eoracana*), 2n = 36, Family: Poaceae. **Minor millets** are Fox Tail Millet (*Setaria italica*), 2n=18, Family: Poaceae; Barn Yard Millet (*Echinochola frumentosa*), 2n=36 and 54, Family: Poaceae; Proso Millet (*Panicum milliaceum*), 2n= 36 and 72, Family: Poaceae; Little Millet (*Panicum sumatranse*), 2n=36, Family: Poaceae; Kodo Millet (*Paspalum scrobiculatum*), 2n=40, Family: Poaceae.

1. Breeding for Oilseed Crops

Vegetable oils are a crucial component of the human diet because they offer an excellent source of energy, a wide range of fatty acids (FAs) required for optimal health, and they aid in the body's digestion of fat-soluble vitamins (such as A, D, E, and K) (Savadi, *et al.*, 2017). Crop yields need to grow to almost 3 billion tonnes by 2050 to fulfil the demands of our expanding global population. Due to the requirement to lower processing costs in the food sector and boost shelf life and oxidative stability of commodity vegetable oils, breeding of high oleic vegetable oils has lately become a crucial component of crop improvement programmes across the world. When carried out properly, plant breeding tackles several genetic and phenotypic goals at once. In order to strategically cross different parent phenotypes and create a better offspring generation, the majority of breeders bring prior knowledge of the parental phenotypes. We may start to build community resources that enable the discovery of genes and quantitative loci that contribute to climate adaptation for future crops using contemporary genomics and using physiological knowledge. It is obvious that traditional breeding efforts must be supplemented with genomic methodologies in order to enhance oil output per unit area, maintain or improve oil quality, and keep up with the growing global demand due to the complex nature of the interplay between genetics and environment. The oilseed-breeding community will be able to broaden the area of investigations in the future thanks to these and innovative high throughput phenotyping techniques in controlled and natural conditions (Attia 2021).

The selection and breeding required to introduce an established oilseed crop to a new location, the quantity and quality of the oil, and the quantity and quality of the meal are the three main goals of oilseed breeding(Knowles 1983). Enhancement of seed and oil production, oil quality according to application (*e.g.*, edible or industrial purposes), breeding of varieties that fit in diverse cropping systems, and breeding of biotic and abiotic stress resistant/tolerant types are major goals in oil crop development(Yadava2012).Oil seeds crops are, Ground Nut (*Arachis hypogcaea*), 2n = 40, Family: Fabaceae; Sunflower (*Helianthus annuus*), 2n=34, Family: Compositae; Sesame (*Sesamum indicum*), 2n=26 Family: Pedaliaceae; Safflower (*Carthamus tinctories*), 2n=24, Family: Asteraceae; Castor (*Ricinus communis*), (2n=20), Family: Euphorbiaceae; Mustard (*Brassica nigra*), (2n=16, 18, 20, 22, 36), Family: Brassiceae (Cruciferae); Linseed (*Linumus itaticimum*), (2n=30),Family: Linaceae.

2. Breeding for Pulses

While lowering the quantities of anti-nutrients in pulse seeds, genetic breeding may aid to maximise beneficial nutrients (Robinson *et al.*, 2019). Despite this, substantial advancements have been achieved in several fields that will help to improve the nutritional value of pulse crops, most notably the modification of the protein content of pea seeds (Domoney *et al.*, 2013).All pulses are considered legumes but not all legumes are considered pulses, the words "legumes" and "pulses" are synonymous. The family Leguminoseae includes pulses. Although they include a limited number of important amino acids like methionine, tryptophan, and cystine, pulses have about 21-25 per cent protein(Singh 2017).

Major breeding goals for grain legumes include early maturity, high yield, broader adaptation, shatter resistance, higher quality, and tolerance to biotic and abiotic stressors. The number of pods per plant, length of pods, number of seeds per pod, and size of seeds are important yield factors (Singh 2017).Pulse crops are Red Gram (*Cajanuscajan*), 2n = 22, Family: Fabaceae; Chickpea (*Cicer arietinum*) 2n = 16, Family: Fabaceae; Soybean (*Glycine max*), 2n = 40 Family: Fabaceae; Green Gram (*Vigna radiata*), 2n = 22, Family: Fabaceae; Black gram (*Vigna mungo*), 2n = 22, Family: Fabaceae; Cowpea (*Vigna unguiculata*), (2n: 22 and 24), Family: Fabaceae; Lab Lab (*Lablab purpureus*), (2n=22, 24), Family: Fabaceae.

3. Breeding for Fibre Crops

Breeding cultivars with higher yields of fibre with better quality is the main breeding strategy for crops used as fibre. In addition to fibre output and quality, other factors including fibre composition, vegetative cycle duration, and resilience to different abiotic and biotic challenges are also thought to be significant (Yadav 2022).

For genomics-assisted breeding, it is essential to improve our capacity to thoroughly interview nucleic-acid-dependent knowledge in the cell. In this regard, molecular markers have demonstrated to be incredibly helpful genetic tools for creating genetic, physical, and phenotypic maps in a range of species used to produce fibre crops (Yadav 2022).

Classification of Plant Breeding Methods in Field Crops

1. Application in Crop Improvement

1. General Methods, Plant introduction, Pure line selection, mass selection, progeny selection, pedigree method, bulk method, back cross method, SSD, clonal selection, heterosis breeding, synthetics and composites (Singh *et al.*, 2007; Singh 2016).
2. Special Methods,Mutation breeding, Polyploidy breeding, transgenic breeding, molecular breeding (Singh *et al.*, 2007; Singh 2016).

3. Population Improvement, Recurrent selection, disruptive selection, diallel selective approaches mating system, biparental mating (Singh *et al.*, 2007; Singh 2016).

2. Hybridization

1. Methods involving hybridization Pedigree, bulk, backcross and SSD Methods: heterosis breeding, and population improvement approaches and molecular breeding (marker aided selection) (Singh *et al.*, 2007; Singh 2016).
2. Methods not involving hybridization Plant Introduction, pureline selection, mass selection, progeny selection, clonal selection, mutation breeding and transgenic breeding (Singh *et al.*, 2007; Singh 2016).

3. Methods of Breeding Autogamous Species

Plant breeding methods that are used for genetic improvement of self pollinated or autogamous species include: Plant Introduction, Pureline selection, Mass selection, Pedigree method, Bulk method, Single seed descent method, Backcross method, Heterosis breeding, Mutation breeding, Polyploidy breeding, Distant hybridization, Transgenic breeding(Singh *et al.*, 2007; Singh 2016). Four breeding approaches, *viz.*, recurrent selection, disruptive selection, diallele selective mating, and biparental mating are used for population improvement (Singh *et al.*, 2007; Singh 2016).

4. Methods or Breeding Allogamous Species

Breeding methods that are used for genetic improvement of cross pollinated orallogamous species include, Plant introduction, Mass and progeny selection, Backcross method, Heterosis breeding, Synthetic breeding, Composite breeding, Polyploidy breeding, Distant hybridization, Transgenic breeding(Singh *et al.*, 2007; Singh 2016). Mutation breeding is rarely used in allogamous species. Three breeding approaches *viz.*, recurrent selection, disruptive mating and biparental mating are used for population improvement(Singh *et al.*, 2007; Singh 2016).

Breeding Methods Commonly Used in Field Crops

1. Plant Introduction

A genotype of a plant or a set of genotypes of plants are introduced into new habitats where they have never been cultivated before. Novel crop species, wild relatives of crop species, or new types of crops currently produced locally can all be introduced. The majority of the materials imported originate from other countries or nations. However, the transfer of crop varieties within a nation from one environment to another is also an introduction (Singh *et al.*, 2007; Singh 2016).

2. Selection

The basis of every crop enhancement is selection. Selection is the process of separating desirable plant kinds from the general population(Singh *et al.*, 2007; Singh 2016).

a. Mass Selection

It is the oldest selecting technique. Since the start of domestication, man has always engaged in mass selection, whether consciously or unconsciously. Mass selection, in its most basic form, is choosing individuals based on phenotypic superiority and combining the seeds for use as planting material the next season(Singh *et al.*, 2007; Hayward 2012; Singh 2016).

b. Pureline Selection

The most popular technique for enhancing self-pollinated crops has been pureline selection. Today's self-pollinated crop types are largely purelines. Numerous improvements to self-pollinating crops can be made through pureline selection. It is used to improve, Local varieties, Old pureline varieties and Introduced varieties(Singh *et al.*, 2007; Singh 2016).

3. Hybridization

Hybridization is the mating or crossing of two plants or lines with different genotypes. When two genotypes, or male and female parents, of a plant are crossed, pollen grains from the male parent are placed on the stigma of the female parent's flower. It is crucial to avoid accidental cross-pollination as well as self-pollination in the female parent's flowers (Singh *et al.*, 2007; Singh 2016; Anushma2021).

a. Combination Breeding

Transferring one or more traits into a single variety from different kinds is the primary goal of combination breeding. These traits might be controlled by polygenes or oligogenes (Singh *et al.*, 2007; Singh 2016; Anushma 2021).

b. Transgressive Breeding

Through transgressive segregation, transgressive breeding seeks to increase yield or the traits that contribute to it. The production of plants in an F_2 generation that are superior to both parents for one or more characteristics is known as transgressive segregation. Such plants are produced by the accumulation of positive or favorable genes from both parents, which must complement one another and are ideally genetically varied, *i.e.*, quite distinct (Singh *et al.*, 2007; Singh 2016; Anushma 2021).

Types of Hybridization

1. Intervarietal and
2. Distant hybridization

Intervarietal Hybridization: In this type of hybridization, two strains, varieties, or races of the same species are the parents. In crop improvement programmes, intervarietal hybridization is the most commonly used (Singh *et al.*, 2007; Singh 2016; Katche2019).

Simple Cross: In a simple cross, two parents are crossed to produce the F_1. The F_1 is selfed to produce F_2 or is used in a backcross programme,

Complex Cross: The hybrid is produced by crossing more than two parents, and it is then utilised in a backcross or to produce F_2. Because this transfer procedure attempts to converging, *i.e.*, bringing together, genes from many parents into a single hybrid, this cross is also known as a convergent cross (Singh *et al.*, 2007; Singh 2016).

Distant Hybridization: Distant hybridization involves sexes from the same genus or from other genus that are not genetically related to one another. When two species of the same genus cross, the process is referred to as interspecific hybridization; however, when they come from two different genera, it is referred to as intergeneric hybridization (Singh *et al.*, 2007; Singh 2016).

4. Pedigree Method

The pedigree approach involves choosing certain plants from F_2 and later generations, then testing their offspring. A record of all parent-offspring is maintained during the whole process. The pedigree record is what it is called. Continued individual plant selection results in progeny that do not exhibit segregation. At this point, progenies are chosen, multilocation testing are carried out, and varieties are made available. The pedigree, which often starts with some distant ancestors, can be characterised as an account of an individual's ancestors. Knowing the relationships between two individuals is helpful for choosing parents and predicting the results of genetic crosses(Singh *et al.*, 2007; Camara and Symonds 2014; Singh 2016).

Another modified pedigree method is the **mass pedigree approach**. This was proposed by Harrington, Up to a favourable season for making desired selections against drought, insects, and diseases, *etc.*, crosses are produced and subsequent generations are mass-produced. From the residual population, individual plants are chosen, their progenies are collected, and their yield and other characteristics are assessed in preliminary yield experiments. Subsequent generations are then carried out using the pedigree technique until the variety is released. It is possible to benefit from both the bulk and pedigree procedures, and a huge number of crossings may be managed at once. It takes a little bit longer, which is a drawback (Kulshrestha1989; Singh *et al.*, 2007; Singh 2016).

5. Bulk Method

Bulk method was first used by Nilsson Ehle in 1908 at Svalof. This method is also known as mass method 'or' Population method of breeding or Opportunity for natural selection.F_2 and later generations are harvested in large quantities as bulk to produce the following generation. Individual plants are chosen and assessed similarly to the pedigree technique of breeding towards the conclusion of the bulking period (after achieving homozygosity).

Modified Bulk Method: Here selection can be practiced in F_2 and F_3 and subsequent generations. There will not be any pedigree record but superior plants are selected bulked and carried forward. In F_4 superior plants are selected and harvested on single plant basis. In F_5 these single plants are studied in progeny rows and best progenies are selected and harvested. In F_6 Preliminary yield trails can be conducted to select best families. In subsequent generations regular trials can be conducted. This modification of the bulk method provides an opportunity for the breeder to exercise his skill and judgement in selection. Further there is no maintenance of pedigree record which is another advantage (Rommens2007; Singh *et al.*, 2007; Singh 2016).

6. Single-Seed-Descent (SSD)

The single-seed-descent (SSD) approach, which is gaining popularity, is a variant of the bulk method. To raise the F_3 generation, one seed is bulked from each of the one to two thousand F_2 plants. Similar to F_3, the following generation is raised by selecting one random seed from each plant in the population and planting it in large quantities. This process is carried upto F_5 or F_6, at which point the plants would have almost reached homozygosity. One to five hundred individual plants are chosen in F_5 or F_6 and their progenies are developed in the next generation. Selection is done mainly among the progenies, and the number of progenies is sufficiently reduced to permit replicated trial in the next generation. Individual plants may be selected only from outstanding families not showing segregation. Thus preliminary yield trials and quality tests begin in F_7 or F_8 and coordinated yield trials in F_8 or F_9(Singh *et al.*, 2007; Singh 2016).

7. Back Cross

Breeders of early 20th century engaged in the development of disease resistant varieties observed that pureline selections with genes for resistance from intra-or interspecific hybridization were inferior to the generally acceptance superior parent in yield or quality characteristic. To overcome this problem, Harlan and Pope 1922 suggested the back cross method by which an undesirable allele at a particular locus is replaced by the desirable allele in otherwise elite variety. The goal is to enhance or correct one or two particular defects in a high yielding variety(Singh *et al.*, 2007; Singh 2016; Begna2021).

Transfer of a Dominant Gene

Let us suppose that a high yielding and widely adapted variety A is susceptible to powdery mildew. Another variety B is resistant to powdery mildew, and that resistance to powdery mildewis dominant to susceptibility. A generalized scheme of the backcross programme for the transfer of powdery mildew resistance from variety B to variety A (Singh *et al.*, 2007; Singh 2016).

Transfer of a Recessive Gene

When powdery mildew resistance is due to a recessive gene, all the backcrosses cannot be made one after the other. After the first backcross, and after every two backcrosses, F_2 must be grown to identify powdery mildew resistant plants. The F_1 and the backcross progenies are not inoculated with powdery mildew because they would be susceptible to powdery mildew. Only the F_2 is tested for powdery mildew resistance. A generalized scheme for the transfer of a recessive gene for powdery mildew resistance has is also well suited to the environment and has other positive traits (Singh *et al.*, 2007; Singh 2016).

8. Recurrent Selection

Breeding schemes similar to recurrent selection were first suggested in 1919 by Hayes and Garber and independently by East and Jones in 1920. Critical data were not given by the above scientist. First detailed description of this type of breeding method was published by Jenkins in 1940 as a result of his experiments with early testing for GCA in maize. The method acquired its name in 1945 when Hull suggested detailed scheme of recurrent selection for SCA, Hull (1952) defined recurrent selection as "Method which involves reselection generation after generation with interbreeding of selects to provide for genetic recombination"(Kutka2011).There are four types of recurrent selections, Simple recurrent selection, Recurrent selection for GCA, Recurrent selection for SCA, Reciprocal recurrent selection(Singh *et al.*, 2007; Singh 2016).

9. Mutation Breeding

Mutation Breeding term coined by Hugo Devries in 1900, derived from latin word. Mutation is the sudden heritable change other than the Mendelian segregation and gene recombination in an organism. A gene change, a chromosomal alteration involving many genes, or a change in the plasmagene can all lead to mutation. Gene or point mutations are mutations caused by changes in the base sequence of genes. Chromosome mutations are mutations caused by changes in chromosome structure or even chromosome number (Singh *et al.*, 2007; Singh 2016).

Molecular Breeding

Molecular breeding, on the other hand, is a more general term; it involves the use of molecular marker data for enhancing the effectiveness of various breeding activities, including planning and execution of breeding programs and improving selection efficiency (Singh and Singh 2015).

1. Marker-Assisted Selection (MAS)

MAS term was first used by Beckmann and Soller (1983), utilizing molecular markers to enhance phenotypic selection for crop development is known as marker-assisted selection (MAS)(Lema2018).Several DNA marker-trait associations have been found as a result of numerous quantitative trait loci (QTLs) mapping investigations for various crop species (Collard and Mackill 2008; Jiang 2013; He *et al.*, 2014; Singh and Singh 2015).

2. Marker Assisted Backcrossing (MAB)

MABC is a precise and efficient technique to introgress a single locus regulating an interesting trait while retaining the important traits of the recurrent parent (Collard and Mackill 2008; Mahmood *et al.*, 2015).In backcross breeding, a useful trait is transferred from a donor parent (DP) into a recurrent parent (RP), which is a superior variety deficient in this trait(Collard and Mackill 2008; Hasan *et al.*, 2015; Singh and Singh 2015).

By utilising markers, MABC shortens the donor segment that contains a target locus, speeds up the recovery of the RP genome during backcrossing, and/ or selects for target loci. The primary goal of MABC is to incorporate a specific gene from sources with inferior agronomic practises (the donor parent) into a proprietary breeding line (the RP)(Hasan *et al.*, 2015; Singh and Singh 2015).

i. Foreground Selection (FS)

Selecting plants using the donor parent's marker allele at the target gene is known as foreground selection. Indirect selection for the target gene/QTL on the basis of linked marker genotype. In order to identify the best plants for backcrossing, foreground selection is also employed to select for reproductive-stage features in the seedling stage. FS is known as positive selection (Takeuchi *et al*2006; Mahmood *et al.*, 2015; Singh and Singh 2015).

ii. Recombinant Selection

Recombinant selection, which occurs at the second level and entails choosing BC progeny with the target gene and recombination events between the target locus and connected flanking markers, is referred to as such. Recombinant selection aims to shrink the donor chromosomal segment's target locus-containing region (*i.e.* size of the introgression). This is significant because crop performance may be severely impacted by several undesired genes related to the target gene from the donor parent, and because the pace of reduction of this donor fragment is slower than for unlinked areas (*i.e.* as linkage drag).Prior to recombinant selection, fine mapping (or high-resolution mapping) is typically necessary. Recombinant selection can also reduce "linkage drag," a "universal nemesis" of plant breeders, by minimizing the donor chromosomal region. (Takeuchi *et al.*, 2006; Mahmood *et al.*, 2015; Singh and Singh 2015).

iii. Background Selection

All genomic areas can be chosen for background selection, with the exception of the target locus, using RP marker alleles. The target locus is chosen based on phenotype. The goal of this selection is to eliminate donor-introduced linkage drag and superfluous genes. It is simple to eliminate the undesirable donor alleles in the same genomic area as the target locus by utilising molecular markers. Using markers that are not related to the target locus, we choose BC progeny with the highest percentage of RP genome at the third level of MABC; this process is known as "background selection." referred to as 'negative selection' selection' (Takeuchi *et al.*, 2006; Mahmood *et al.*, 2015; Singh and Singh 2015).

3. Gene Pyramiding

Pyramiding entails stacking multiple genes leading to the simultaneous expression of more than one gene in a variety to develop durable resistance expression. Gene pyramiding is gaining considerable importance as it would improve the efficiency of plant breeding leading to the development of genetic stocks and precise development of broad spectrum resistance capabilities. The success of gene pyramiding depends upon several critical factors, including the number of genes to be transferred, the distance between the target genes and flanking markers, the number of genotype selected in each breeding generation, the nature of germplasm *etc.*(Joshi and Nayak2010; Singh and Singh 2015).

4. Marker-Assisted Recurrent Selection (MARS)

Marker-assisted Recurrent Selection (MARS) is a recurrent selection method that uses molecular markers to identify and select several genomic areas that are important in the expression of complex characteristics in order to construct the optimal genotype within a single population or across related populations (Ribaut *et al.*, 2010). MARS is a programme that enables genotypic selection and intercrossing between the chosen individuals throughout the same crop season for one cycle of selection (Jiang *et al.*, 2007). MARS might therefore improve the effectiveness of recurrent selection and speed up the process, especially by assisting in the integration of several advantageous genes or QTLs from various sources through recurrent selection based on multi-parent populations (Gazal *et al.*, 2015; Singh and Singh 2015; Gokidi *et al.*, 2016).

5. Early Generation Marker-Assisted Selection

Early generation MAS is very advantageous because plants with undesired gene combinations may be excluded. Early generation MAS markers can be utilised at any step of a normal plant breeding programme. This enables breeders to concentrate their efforts on fewer high-priority lineages in succeeding generations. The highest effectiveness of MAS occurs in the early generations when the linkage between the marker and the chosen QTL is not very tight due to the rising likelihood of recombination between the marker and QTL. The cost of genotyping a greater number of plants is the main drawback of using MAS in early generations (Collard and Mackill2008; Singh and Singh 2015).

6. Combined Marker-Assisted Selection

Phenotypic screening can be carefully integrated with MAS in a number of situations. In order to optimise genetic gain, "combined MAS" may be preferable than phenotypic screening or MAS alone in the first instance. This strategy could be used when there are still undiscovered QTLs influencing a characteristic or when several QTLs need to be altered (Moreau *et al.*, 2004; Singh *et al.*, 2023). When using large population sizes and low trait heritabilities, simulation studies show that this strategy is more effective than phenotypic screening alone. In reality, phenotypic selection will be applied at various phases of the breeding procedure, which encompasses all MAS schemes. This will be required to both select for characteristics or genes for which the map position is unclear and to confirm the MAS result(Bohn *et al.*, 2001; Singh and Singh 2015; Singh *et al.*, 2023).

7. Advanced Backcross-QTL (AB-QTL)

The simultaneous identification and transfer of important QTLs from unadapted germplasm into top breeding lines using advanced backcross QTL (AB-QTL) study is an effective way to use wild species. In order to find and incorporate advantageous QTLs from wild species into elite breeding lines, the AB-QTL technique, which encourages deferring QTL analysis until BC_2 or BC_3, was developed. The majority of the important crop species have AB-QTL populations, and the examples shown here show that this strategy has significantly aided in releasing advantageous alleles from the wild species parents(Singh and Singh 2015; Wang and Chee 2010).

8. Linkage Disequilibrium Based Mapping

Linkage disequilibrium, also known as gametic phase disequilibrium, gauges how non-randomly alleles are associated at various loci. LD can be measured by D, D2 and r^2 (Zhu *et al.*, 2008; Singh and Singh 2015).

(a) Association Mapping

By looking at the marker-trait relationships that may be explained by the degree of linkage disequilibrium between markers and functional polymorphisms across a set of varied germplasm, association mapping discovers quantitative trait loci (QTLs). Numerous quantitative trait loci (QTLs), their interactions, the environment, and the combination between QTL and environment all influence the phenotypic variance of many complex variables of agricultural or evolutionary importance (Zhu *et al.*, 2008; Singh and Singh 2015).

(b) Genome-Wide Association Studies (GWAS)

GWAS have grown to be a potent and popular approach to studying complex traits. By assessing genome-wide SNPs in a gathered population, GWAS analyse complex traits. In several significant crop species, genome-wide association research has examined features that are crucial for agriculture .Other kinds of traits have also been investigated using genome-wide association analyses. Genome-wide association studies are used to both find new associations with important characteristics and to confirm loci found using other techniques (Singh and Singh 2015; Tibbs *et al*2021).

9. Genomic Selection

By using genome-wide markers in breeding programmes, the predictive strategy known as genomic selection (GS) aims to accelerate the rate of genetic gain per unit of time and shorten the generation interval. It has become a useful technique for enhancing complicated features that are governed by several genes with modest impacts. The breeding value of potential genotypes for selection may be predicted using GS. Despite being the most widely used technique and frequently used interchangeably with genomic prediction is genomic best linear unbiased prediction (GBLUP) (Cappetta *et al.*, 2020; Montesinos *et al.*, 2021).Genomic selection (GS), proposed by Bernardo and Meuwissen *et al.* In recent years, breeding programmes have begun to use genomic selection (GS) as a potent technique for selection. The breeding technique known as genomic selection (GS) has been used effectively in the breeding of animals and is now being used in the breeding of plants. By quickly selecting genotypes with superior traits, GS promises to decrease breeding operations. Genome estimated breeding values (GEBVs), which have a significant potential to increase the selection efficiency, are used in GS to guide genotype selection (Singh and Singh 2015; Krishnappa *et al.*, 2021).

New Breeding Techniques

New breeding techniques (NBTs) make specific changes within plant DNA in order to change its traits, and these modifications can vary in scale from altering single base, to inserting or removing one or more genes.

The following plant-breeding techniques can currently be considered as the main NBTs:

Site-Directed Nucleases (SDN) (including ZFN and CRISPR systems), Oligonucleotide Directed Mutagenesis (ODM), Cisgenesis, RNA-dependent DNA methylation (RdDM), Grafting (non-GM scion on GM rootstock), Reverse breeding, Agro-infiltration(Van *et al.*, 2019).

Cutting and modifying the genome during the repair process (this tools are used to achieve this: Zinc finger nuclease; TALENs, and CRISPR/Cas Tools). Genome editing to introduce changes to just a few base pairs (using a technique called 'oligonucleotide-directed mutagenesis' (ODM). Transferring a gene from an identical or closely related species (cisgenesis). Adding in a reshuffled set of regulatory instructions from same species (intragenesis).Deploying processes that alter gene activity without altering the DNA itself (epigenetic methods). Grafting of unaltered plant onto a genetically modified rootstock(Van *et al.*, 2019).

1. Genome Editing

Using DNA repair mechanisms, genome editing and gene editing are methods for genome engineering that introduce site-specific alterations into genomic DNA. Because it only affects one gene, gene editing is distinct from genome editing. Four major mechanisms of site-specific genome editing, meganucleases (MegNs), zinc finger nucleases (ZFNs), transcription activator-like effector nuclease (TALENs), and clustered regularly interspaced short palindromic repeats (CRISPR)/CRISPR-associated protein 9 (Cas9) (CRISPR/Cas-9) (Khalil 2020).

2. Meganucleases (MegNs)

Meganucleases (MegNs) are endodeoxy ribonucleases that exist naturally and are present in all microbial life forms as well as in eukaryotic mitochondria and chloroplasts. Self-splicing elements frequently include the genes that encode MegNs (Khalil 2020).

3. Zinc Finger Nucleases (ZFNs)

The advent of zinc finger nucleases marked the beginning of genome editing technologies (ZFNs). Zinc finger nucleases are bespoke site-specific restriction enzymes produced by genetic engineering. The transcription factors known as zinc fingers each identify three to four bases. One of the DNA binding motifs present in the DNA binding domain of many eukaryotic transcription factors, which is involved in DNA identification, is the zinc finger. A nuclease (typically from the bacterial restriction enzyme FokI) is found in the second domain.ZFN was enticing and had great potential; it was employed in many living things or cultured cells (Carroll 2011; Khalil 2020).

4. TALENs

In that the proteins must be modified for each targeted DNA sequence, TALENs are conceptually comparable to ZFNs and MegNs. Both the TALENs and ZFNs are modular and contain inherent DNA-binding specificities. Additionally, the application of TALENs in plant genome editing has been more fruitful. TALENs have an advantage over ZFNs in that their DNA-binding domain only identifies one nucleotide per module, as opposed to the initial single zinc finger domains' 3 bps recognition (Zhang *et al.*, 2013; Khalil 2020).

5. CRISPR/Cas System

The most current platform for genome editing is the CRISPR/Cas system. The system, which is referred to as third generation genome editing tools, was developed in 2013. The 1980s had seen finding of the clustered regularly interspaced short palindromic repeats, often known as "short regularly spaced repeats."It was proposed to use the term CRISPR, and a collection of genes around the CRISPR locus known as the "CRISPR-associated system" or Cas was identified(Jansen 2002).The DNA recognition of the CRISPR/Cas system is based on RNA-DNA interactions, in contrast to the two prior technologies (ZFN and TALEN), in which the recognition of the DNA site was based on sequence recognition by artificial proteins necessitating contact between protein and

DNA. Compared to ZFNs and TALENs, this has a number of benefits.When Cas9 nuclease cleaves DNA, it can produce a double strand break (DSB) when employing the wild-type enzyme or singlestrand break (SSB) when utilising recombinant Cas9 variations known as nickases. Using CRISPR/Cas systems, immunity develops in three phases. 1. adaptation, 2. CRISPR processing and transcription (crRNA generation), 3. Interference (Vasebi and Khakvar 2014). The usage of the CRISPR/Cas9 system might lead to non-specific and unexpected genomic alterations (off-target impact), which is one of its downsides. Future research must address the off-target impact if CRISPR/Cas9 is to live up to its potential as a potent genome editing technique (Mahas and Mahfouz 2018).

6. Cisgenesis

Since genes of interest are separated from a species that may be used for traditional crosses and transferred, maintaining its "natural" form, cisgenic plants may resemble plants obtained from traditional breeding and share the same genetic pool with them. Shouten initially suggested the concept of cisgenesis in 2006. Crops modified using genes extracted only from sexually compatible plants, including gene introns and regulatory areas like promoters and terminators, are considered the end product of cisgenic techniques according to a commonly recognised definition. It takes a productive worker with exceptional expertise as well as additional time. The resistance gene HcrVf2 was inserted into the apple cultivar Gala to produce the first cisgenic plant for scab resistance (Schouten *et al.*, 2006; Mohapatra and Sahoo 2020).

7. Reverse Breeding

By directly developing parental lines for any heterozygous plant, reverse breeding (RB) is a cutting-edge plant breeding approach created to accomplish one of the most sought-after goals of plant breeding. Through controlled meiosis, RB produces homozygous parental lines that are precisely complementary. The major goal is to comprehend the method for producing homozygous parental lines, which is based on preventing meiotic crossing over in the chosen heterozygote to reduce genetic recombination. These plants produce male or female spores, both of which contain non-recombinant parental chromosomal combinations that may be grown in vitro to produce homozygous doubled haploid plants (DHs).Reverse breeding is required to preserve hybrid stability, improve the genetic makeup of the parental lines, and increase hybrid performance. For uncharacterized heterozygotes, breeding lines must be established, from homozygous parental lines to create a highly heterozygous plant (Lusser *et al.*, 2011; Singh 2018; Calvo-Baltanás *et al.*, 2021).

Speed Breeding

All plant scientists were inspired by NASA's efforts in the early 1980s. The phrase "speed breeding," which refers to a collection of techniques used to quicken wheat breeding, was first used in 2003 by researchers at the University of Queensland. Time-saving techniques for accelerating plant breeding include speed breeding. The majority of plant species have bottlenecks in their applied research and breeding programmes, necessitating the development of technology to quicken generational turnover and plant growth (Watson *et al.*, 2018; Ahmar *et al.*, 2020). For a number of crops, speed breeding procedures are now being developed. Contrary to doubled haploid (DH) technology, which produces haploid embryos to generate entirely homozygous lines, speed breeding is suited for a variety of germplasm and doesn't need for specialised in vitro culturing equipment (Slama-Ayed *et al.*, 2019; Ahmar *et al.*, 2020). The concept behind speed breeding is to employ the best conditions for light, temperature, and day length regulation (22 hours of light, 22 degrees Celsius during the day, and 17 degrees Celsius at night), as well as annual seed harvesting, to accelerate photosynthesis and prompt early blooming. A significant factor in controlling flowering is light intensity and wavelength.(Watson *et al.*, 2018; Chiurugwi *et al.*, 2019; Ahmar *et al.*, 2020).

Under regulated lighting conditions, Croser *et al.* developed early- and late-flowering genotypes for beans, peas, chickpeas, faba and lupins using metal halide and enhanced blue and far-red LED lights. These species had a favourable connection with the declining red:far red-red ratio (R:FR) (Croser *et al.*, 2016).

Speed breeding was once thought to reduce generation time by lengthening photoperiods, although some crop species, including radish (*Raphanus sativus*), pepper (*Capsicum annum*), and sunflower (*Helianthus annuus*), reacted well to longer days (Sysoeva *et al.*, 2010; Ahmar *et al.*, 2020).Short-day crop breeding has been slowed down due to the need for blooming. However, as part of a project supported by the Bill and Melinda Gates Foundation, Lee Hickey and his research team recently collaborated with ICRISAT to develop protocols for short-day crops like sorghum, millet, and pigeon pea. Many smallholder farmers depend on sorghum, millet, and pigeon peas, thus developing protocols with their needs in mind will have a huge impact on subsistence farming worldwide. Speed breeding can hasten vegetative development in staple food crops like rice (*Oryza sativa*) and maize (*Zea mays*), which need shorter photoperiods to begin the reproductive stages. Speed breeding makes it feasible to produce consecutive generations of enhanced crops for field testing using SSD at a lower cost than creating DHs. Additionally, MAS of top hybrid lines followed by speed breeding is advantageous for gene insertion (common haplotypes) of diverse phenotypes (Collard *et al.*, 2017; Hickey *et al.*, 2017; Wolter *et al.*, 2019; Ahmar *et al.*, 2020).

Conclusion

In conclusion, new developments in plant breeding and genomics have aided in the creation of cultivars that are both qualitatively and quantitatively better. Amazing agricultural results have been achieved using both conventional approaches and novel agronomic strategies. However, the governmental sector, private businesses, and extension agents must all invest together for sustainable crop improvement to assure global food security.

REFERENCES

1. Stamp, P., and Visser, R. (2012). The twenty-first century, the century of plant breeding. *Euphytica, 186*(3), 585-591.
2. Cattivelli, L., Rizza, F., Badeck, F. W., Mazzucotelli, E., Mastrangelo, A. M., Francia, E., and Stanca, A. M. (2008). Drought tolerance improvement in crop plants: an integrated view from breeding to genomics. *Field crops research, 105*(1-2), 1-14.
3. Singh, R. J., and Jauhar, P. P. (2006). *Genetic Resources, Chromosome Engineering, and Crop Improvement: Cereals, Volume 2*. CRC press.
4. Borrill, P. (2020). Blurring the boundaries between cereal crops and model plants. *New Phytologist, 228*(6), 1721-1727.
5. Sarwar, M. H., Sarwar, M. F., Sarwar, M., Qadri, N. A., and Moghal, S. (2013). The importance of cereals (Poaceae: Gramineae) nutrition in human health: A review. *Journal of cereals and oilseeds, 4*(3), 32-35.
6. Borrill, P. (2020). Blurring the boundaries between cereal crops and model plants. *New Phytologist, 228*(6), 1721-1727.
7. Singh, B. D., and Ashok Kumar Singh. "Marker-assisted plant breeding: principles and practices." (2015): 978-81.
8. Singh, B. D. (2016). Plant Breeding: Principles and Methods. Kalyani publishers.
9. Singh, P., Singh, P., and Singh, P. (2007). *Essentials of plant breeding*. Kalyani publishers.
10. Loskutov, I. G. (2021). Advances in cereal crops breeding. *Plants, 10*(8), 1705.
11. Savadi, S., Lambani, N., Kashyap, P. L., and Bisht, D. S. (2017). Genetic engineering approaches to enhance oil content in oilseed crops. *Plant Growth Regulation, 83*, 207-222.
12. Attia, Z., Pogoda, C. S., Reinert, S., Kane, N. C., and Hulke, B. S. (2021). Breeding for sustainable oilseed crop yield and quality in a changing climate. *Theoretical and Applied Genetics, 134*(6), 1817-1827.
13. Knowles, P. F. (1983). Genetics and breeding of oilseed crops. *Economic Botany, 37*(4), 423-433.
14. Yadava, D. K., Vasudev, S., Singh, N., Mohapatra, T., and Prabhu, K. V. (2012). Breeding major oil crops: Present status and future research needs. *Technological innovations in major world oil crops, volume 1: breeding*, 17-51.

15. Robinson, G. H. J., Balk, J., and Domoney, C. (2019). Improving pulse crops as a source of protein, starch and micronutrients. *Nutrition bulletin, 44*(3), 202-215.
16. Domoney, C., Knox, M., Moreau, C., Ambrose, M., Palmer, S., Smith, P., and Ellis, N. (2013). Exploiting a fast neutron mutant genetic resource in Pisumsativum (pea) for functional genomics. *Functional Plant Biology, 40*(12), 1261-1270.
17. Singh, N. (2017). Pulses: an overview. *Journal of Food Science and Technology, 54*, 853-857.
18. Mafongoya, P. L., and Sileshi, G. W. (2020). Indices to identify and quantify ecosystem services in sustainable food systems. In *The Role of Ecosystem Services in Sustainable Food Systems* (pp. 43-71). Academic Press.
19. Yadav, S., Singh, V. K., Bhoyar, P. I., Sharma, V., Rehsawla, R., and Kumar, R. (2022). Emerging Technologies in Plant Breeding for Fibre Crops, Cotton, and Sunn Hemp. *Technologies in Plant Biotechnology and Breeding of Field Crops*, 151-180.
20. Hayward, M. D., Bosemark, N. O., and Romagosa, T. (Eds.). (2012). *Plant breeding: principles and prospects*. Springer Science and Business Media.
21. Anushma, P. L., Dhanyasree, K., and Rafeekher, M. (2021). Wide hybridization for fruit crop improvement: A review. *International Journal of Chemical Studies, 9*(1), 769-773.
22. Katche, E., Quezada-Martinez, D., Katche, E. I., Vasquez-Teuber, P., and Mason, A. S. (2019). Interspecific hybridization for Brassica crop improvement. *Crop Breeding, Genetics and Genomics, 1*(1).
23. Camara, M. D., and Symonds, J. E. (2014). Genetic improvement of New Zealand aquaculture species: programmes, progress and prospects. *New Zealand Journal of Marine and Freshwater Research, 48*(3), 466-491.
24. Kulshrestha, V. P. (1989). A modified pedigree method of selection. *Theoretical and applied genetics, 78*, 173-176.
25. Rommens, C. M., Haring, M. A., Swords, K., Davies, H. V., and Belknap, W. R. (2007). The intragenic approach as a new extension to traditional plant breeding. *Trends in plant science, 12*(9), 397-403.
26. Begna, T. (2021). Conventional Breeding Methods Widely used to Improve Self-Pollinated Crops. *International Journal of Research Studies in Agricultural Sciences* Volume 7, 2454–6224.
27. Kutka, F. (2011). Open-pollinated vs. hybrid maize cultivars. *Sustainability, 3*(9), 1531-1554.
28. Joshi, R. K., and Nayak, S. (2010). Gene pyramiding-A broad spectrum technique for developing durable stress resistance in crops. *Biotechnol. Mol. Biol. Rev, 5*(3), 51-60.

29. Lema, M. (2018). Marker assisted selection in comparison to conventional plant breeding. *Agric Res Technol, 14*, 555914.
30. He, J., Zhao, X., Laroche, A., Lu, Z. X., Liu, H., and Li, Z. (2014). Genotyping-by-sequencing (GBS), an ultimate marker-assisted selection (MAS) tool to accelerate plant breeding. *Frontiers in plant science, 5*, 484.
31. Collard, B. C., and Mackill, D. J. (2008). Marker-assisted selection: an approach for precision plant breeding in the twenty-first century. *Philosophical Transactions of the Royal Society B: Biological Sciences, 363*(1491), 557-572.
32. Jiang, G. L. (2013). Molecular markers and marker-assisted breeding in plants. *Plant breeding from laboratories to fields, 3*, 45-83.
33. Hasan, M. M., Rafii, M. Y., Ismail, M. R., Mahmood, M., Rahim, H. A., Alam, M. A and Latif, M. A. (2015). Marker-assisted backcrossing: a useful method for rice improvement. *Biotechnology and Biotechnological Equipment, 29*(2), 237-254.
34. Takeuchi, Y., Ebitani, T., Yamamoto, T., Sato, H.,Ohta, H., Hirabayashi, H and Yano, M. (2006). Development of isogenic lines of rice cultivar Koshihikari with early and late heading by marker-assisted selection. *Breeding Science, 56*(4), 405-413.
35. Joshi, R. K., and Nayak, S. (2010). Gene pyramiding-A broad spectrum technique for developing durable stress resistance in crops. *Biotechnol. Mol. Biol. Rev, 5*(3), 51-60.
36. Gokidi, Yugandhar.,Bhanu, A. N., and Singh, M. N. (2016). Marker assisted recurrent selection: an overview. *Adv Life Sci, 5*(17), 6493-9.
37. Ribaut, J. M., De Vicente, M. C., and Delannay, X. (2010). Molecular breeding in developing countries: challenges and perspectives. *Current opinion in plant biology, 13*(2), 213-218.
38. Jiang, G. L., Shi, J., and Ward, R. W. (2007). QTL analysis of resistance to Fusarium head blight in the novel wheat germplasm CJ 9306. I. Resistance to fungal spread. *Theoretical and Applied Genetics, 116*, 3-13.
39. Gazal, A., Dar, Z. A., Lone, A. A., Abidi, I., and Ali, G. (2015). Molecular breeding for resilience in maize-a review. *Journal of Applied and Natural Science, 7*(2), 1057-1063.
40. Singh, N. K., Joshi, A., Sahoo, S., Tufchi, M., and Rakshit, S. (2023). Molecular breeding for improving yield in maize: Recent advances and future perspectives. *QTL Mapping in Crop Improvement*, 75-99.
41. Moreau, L., Charcosset, A., and Gallais, A. (2004). Experimental evaluation of several cycles of marker-assisted selection in maize. *Euphytica, 137*(1), 111-118.

42. Bohn, M., Groh, S., Khairallah, M. M., Hoisington, D. A., Utz, H. F., and Melchinger, A. E. (2001). Re-evaluation of the prospects of marker-assisted selection for improving insect resistance against *Diatraea spp.* in tropical maize by cross validation and independent validation. *Theoretical and Applied Genetics, 103.*
43. Wang, B., and Chee, P. W. (2010). Application of advanced backcross quantitative trait locus (QTL) analysis in crop improvement. *Journal of plant Breeding and crop Science,* 2(8), 221-232.
44. Van Vu, T., Sung, Y. W., Kim, J., Doan, D. T. H., Tran, M. T., and Kim, J. Y. (2019). Challenges and perspectives in homology-directed gene targeting in monocot plants. *Rice, 12*(1), 1-29.
45. Khalil, A. M. (2020). The genome editing revolution. *Journal of genetic engineering and biotechnology, 18*(1), 1-16.
46. Carroll, D. (2011). Genome engineering with zinc-finger nucleases. *Genetics, 188*(4), 773-782.
47. Zhang, Y., Zhang, F., Li, X., Baller, J. A., Qi, Y., Starker, C. G., and Voytas, D. F. (2013). Transcription activator-like effector nucleases enable efficient plant genome engineering. *Plant physiology, 161*(1), 20-27.
48. Jansen, R., Embden, J. D. V., Gaastra, W., and Schouls, L. M. (2002). Identification of genes that are associated with DNA repeats in prokaryotes. *Molecular microbiology, 43*(6), 1565-1575.
49. Vasebi, Y., and Khakvar, R. (2014). CRISPR-Cas: the effective immune systems in the prokaryotes. *International Journal of Molecular and Clinical Microbiology,* 4(1), 334-344.
50. Mahas, A., and Mahfouz, M. (2018). Engineering virus resistance via CRISPR–Cas systems. *Current opinion in virology, 32,* 1-8.
51. Schouten, H. J., Krens, F. A., and Jacobsen, E. (2006). Cisgenic plants are similar to traditionally bred plants: international regulations for genetically modified organisms should be altered to exempt cisgenesis. *EMBO reports,* 7(8), 750-753.
52. Mohapatra, U., and Sahoo, J. P. Cisgenesis: Can it be an Alternative to Transgenesis?. *Trends in plant science,* 6(6), 262-267.
53. Lusser, M., Parisi, C., Plan, D., and Rodríguez-Cerezo, E. (2011). New plant breeding techniques. *Stateof-the-art and prospects for commercial development.* (= *JRC Scientific and Technical Reports/EUR 24760 EN*).
54. Singh, N. K. (2018). Reverse breeding: Accelerating innovation in plant breeding. *Journal of Pharmacognosy and Phytochemistry,* 7(1S), 1811-1813.
55. Calvo-Baltanás, V., Wang, J., and Chae, E. (2021). Hybrid incompatibility of the plant immune system: an opposite force to heterosis equilibrating hybrid performances. *Frontiers in Plant Science, 11,* 576796.

56. Watson, A., Ghosh, S., Williams, M. J., Cuddy, W. S., Simmonds, J., Rey, M. D.,. and Hickey, L. T. (2018). Speed breeding is a powerful tool to accelerate crop research and breeding. *Nature plants*, 4(1), 23-29.

57. Ahmar, S., Gill, R. A., Jung, K. H., Faheem, A., Qasim, M. U., Mubeen, M., and Zhou, W. (2020). Conventional and molecular techniques from simple breeding to speed breeding in crop plants: recent advances and future outlook. *International journal of molecular sciences, 21*(7), 2590.

58. Slama-Ayed, O., Bouhaouel, I., Ayed, S., De Buyser, J., Picard, E., and Amara, H. S. (2019). Efficiency of three haplomethods in durum wheat (*Triticumturgidum* subsp. *durum*Desf.): Isolated microspore culture, gynogenesis and wheat× maize crosses. *Czech Journal of Genetics and Plant Breeding, 55*(3), 101-109.

59. Chiurugwi, T., Kemp, S., Powell, W., and Hickey, L. T. (2019). Speed breeding orphan crops. *Theoretical and Applied Genetics, 132*, 607-616.

60. Watson, A., Ghosh, S., Williams, M. J., Cuddy, W. S., Simmonds, J., Rey, M. D., and Hickey, L. T. (2018). Speed breeding is a powerful tool to accelerate crop research and breeding. *Nature plants*, 4(1), 23-29.

61. Croser, J. S., Pazos-Navarro, M., Bennett, R. G., Tschirren, S., Edwards, K., Erskine, W., and Ribalta, F. M. (2016). Time to flowering of temperate pulses in vivo and generation turnover in vivo–in vitro of narrow-leaf lupin accelerated by low red to far-red ratio and high intensity in the far-red region. *Plant Cell, Tissue and Organ Culture* (*PCTOC*), *127*, 591-599.

62. Sysoeva, M. I., Markovskaya, E. F., and Shibaeva, T. G. (2010). Plants under continuous light: a review. *Plant stress*, 4(1), 5-17.

63. Collard, B. C., Beredo, J. C., Lenaerts, B., Mendoza, R., Santelices, R., Lopena, V., and Islam, M. R. (2017). Revisiting rice breeding methods–evaluating the use of rapid generation advance (RGA) for routine rice breeding. *Plant Production Science, 20*(4), 337-352.

64. Hickey, L. T., Germán, S. E., Pereyra, S. A., Diaz, J. E., Ziems, L. A., Fowler, R. A., and Dieters, M. J. (2017). Speed breeding for multiple disease resistance in barley. *Euphytica, 213*, 1-14.

65. Wolter, F., Schindele, P., and Puchta, H. (2019). Plant breeding at the speed of light: the power of CRISPR/Cas to generate directed genetic diversity at multiple sites. *BMC plant biology, 19*(1), 1-8.

66. Zhu, C., Gore, M., Buckler, E. S., and Yu, J. (2008). Status and prospects of association mapping in plants. *The plant genome, 1*(1).

67. Tibbs Cortes, L., Zhang, Z., and Yu, J. (2021). Status and prospects of genome wide association studies in plants. *The plant genome, 14*(1), e20077.

68. Cappetta, E., Andolfo, G., Di Matteo, A., Barone, A., Frusciante, L., and Ercolano, M. R. (2020). Accelerating tomato breeding by exploiting genomic selection approaches. *Plants, 9*(9), 1236.

69. Montesinos-López, O. A., Montesinos-López, A., Pérez-Rodríguez, P., Barrón-López, J. A., Martini, J. W., Fajardo-Flores, S. B.,. and Crossa, J. (2021). A review of deep learning applications for genomic selection. *BMC genomics, 22,* 1-23.

70. Krishnappa, G., Savadi, S., Tyagi, B. S., Singh, S. K., Mamrutha, H. M., Kumar, S., and Singh, G. P. (2021). Integrated genomic selection for rapid improvement of crops. *Genomics, 113*(3), 1070-1086.

71. https: //www.britannica.com/science/plant-breeding/Breeding-self-pollinated-species

72. Collard, B. C., and Mackill, D. J. (2008). Marker-assisted selection: an approach for precision plant breeding in the twenty-first century. *Philosophical Transactions of the Royal Society B: Biological Sciences, 363*(1491), 557-572.

73. Hospital, F. (2005). Selection in backcross programmes. *Philosophical Transactions of the Royal Society B: Biological Sciences, 360*(1459), 1503-1511.

74. Hospital, F. (2001). Size of donor chromosome segments around introgressed loci and reduction of linkage drag in marker-assisted backcross programs. *Genetics, 158*(3), 1363-1379.

Chapter 8

Crop-Specific Breeding Methods (Horticultural Crops)

B.C. Akhilraj[1] *and J. Suresh*[2]

1Ph.D Scholar, Department of Plantation, Spices, Medicinal and Aromatic Crops, HC and RI, Tamil Nadu Agricultural University, Coimbatore, Tamil Nadu
e-mail: bcakhilraj@gmail.com
[2]*Professor and Head, Coconut Research Station, Tamil Nadu Agricultural University, Aliyar Nagar, Pollachi, Tamil Nadu*
e-mail: sureshpreet2006@yahoo.co.in

Introduction

Horticulture, a significant aspect of agriculture, is the management, production, and sale of plants such as fruits, vegetables, flowers, decorative plants, medicinal plants, and fragrant plants. Horticulture has demonstrated its significance in a number of innovative areas, including bettering land management, encouraging crop diversity, creating jobs, and ensuring the public's nutritional security (Sonah *et al.,* 2011). Vegetative propagation is used for a lot of horticultural crops since it allows for the fixing and propagation of advantageous genetic combinations. Plant breeding is an art, science, and technology that aims to improve the genetic make-up of plants for greater economic and societal usage. Plant breeding aims to enhance agricultural plant traits so that they become more appealing from an economic and agronomic standpoint. Depending on the crop in question, the particular goals vary greatly. Along the way of human civilisation, efforts were made to advance plant breeding for societal benefit. The main objective is to feed the world's growing population. Additionally, because it is a need in today's market, quality enhancement is one of the biggest obstacles encountered by those involved in plant breeding. Both before and during the Green Revolution, diverse agricultural plants have evolved a variety of kinds via the use of classical genetics. The field of molecular biology has advanced, which has sharpened the fundamental breeding tools and increased the prospects of confidence to serve the civilisation. New plant varieties are successfully bred through a number of fundamental processes that are crucial to notice (Prohens, *et al.*, 2008).

A horticulture crop's top priority is quality. *De gustibus non estdisputandum*, or "there is no disputing taste," is undoubtedly true, but the reality is that many horticultural crop cultivars with distinctive features are so exceptional, adored, and valued that they serve as the foundation for whole businesses. Examples include the "Williams" (or "Bartlett") pear, which is nearly the sole cultivar used for processing and dominates the production of fresh fruit, "Hayward" kiwifruit, "Kerman" pistachio, "Smooth Cayenne" pineapple (for processing), "Clementine" mandarin, and "Dwarf Cavendish" banana. The goals of genetic improvement differ for crops grown for horticulture and agronomy. Agronomic crops are transformed into commodities, where the product may be exchanged (Kumar, 2006).

The goal of breeding is to increase yield, which is frequently achieved through improving resistance to biotic and abiotic stress. For instance, boosting yield stability under large populations is the basis for improving hybrid maize yields. Because customers choose between various cultivars and alternative crop species as well as what they will consume, breeding objectives for horticulture crops must be consumer-directed. There are several instances of extensive breeding programmes that failed to get the support of growers because they were unable to compete in the market on the basis of quality. For instance, customers don't care about the disease resistance of food crops; instead, they buy based on what they see and continue to buy based on what they taste. Thus, the main breeding goal must be to produce an exceptional quality rather than yield in and of itself. It must be emphasised that consumer-directed qualities, particularly those related to quality, are frequently intractable by non-genetic methods, in contrast to grower-directed traits, which may frequently be handled by non-genetic means (Bliss, 2010).

The impact of each breeder's creativity in suggesting novel improvements has a further substantial impact on plant breeding. The following list of four significant horticultural breeding developments over the past fifty years is meant to illustrate this idea. These outstanding accomplishments are the result of individual inspiration and study. These accomplishments highlight the influence of creativity, aptitude, and the art and science of plant breeding on the path of horticulture in the future.

The primary steps in plant breeding are as follows:

1. Creation of variety
2. Selection of elite genotypes
3. Evaluation
4. Release
5. Multiplication
6. Distribution of the new variety

Breeders of horticultural crops tend to focus more on quality attributes than yield, which is only the biomass generated in the field. Flavor, colour, shape, size, degree of damage, nutritional content, and characteristics that promote a higher perception of food safety or environmental sustainability are all examples of quality.

Different categories can be used to categorise various quality traits:

Type of Traits	*Description*	*Examples*
Morphological	Appearance of the produce	Fruit size, shape, colour
Organoleptic	Palatability of the produce	Taste, aroma, juiciness, softness
Nutritional	Value of the produce for human or animal nutrition	Protein content, oil content, vitamin content,mineral content
Consumer preference	Increases the marketability of the produce	Keeping quality, storability

Plant Breeding Methods

The reproductive behaviour of the particular crop is always used to find the best breeding method to adapt to a given circumstance. Crop plants are categorised into three classes based on how they reproduce and pollinate (Janick, 2004). These divisions include:

1. Self-pollinated
2. Cross-pollinated
3. Vegetatively (asexually) propagated

1. Self-Pollinated Species

Autogamous or self-pollinating species are those that reproduce by themselves. These plants self-pollinate in order to produce their seeds (autogamy). Self-pollinated organisms are hence also referred to as inbreeders. Numerous traits of plants that encourage self-fertilization include homogamy, cleistogamy, chasmogamy, bisexuality, *etc.* The following are some crucial traits of autogamous species:

1. They regularly engage in self-pollination

2. They are genuine breeding type, being homozygous.
3. In general, lack detrimental genes that are passed down through recessive inheritance because these genes are removed by inbreeding and gene fixation.
4. Self-pollination prevents novel gene combinations in autogamous species.

Techniques Used in Breeding Autogamous Species

1. Plant introduction
2. Pure line selection
3. Mass selection
4. Pedigree method
5. Bulk method
6. Single seed descent method
7. Backcross method
8. Heterosis breeding
9. Mutation breeding
10. Polyploidy breeding
11. Distant hybridisation
12. Transgenic breeding

Self-pollinated crops utilise recurrent selection, disruptive selection, diallel selective mating, and biparental mating to boost their population.

2. Cross-Pollinated Species

Cross-pollinating species, commonly referred to as allogamous species or out breeders, generate seed by cross-pollination (allogamy). Numerous floral characteristics, including as dichogamy, monoecy, dioecy, heterostyly, herkogamy, self-incompatibility, and male sterility, encourage cross-pollination. Some key characteristics of our breeders include:

1. They randomly mate, for one. Each genotype in such a population (panmictic) has an equal probability of mating with every other genotype.
2. Each person has a heterozygous genetic composition.
3. People contain harmful recessive genes, but the influence of dominant genes masks them.
4. Out breeders have severe inbreeding depression when they try to conceive.
5. New gene pairings are made possible through cross-pollination.

Methods Used in Breeding Allogamous Species

1. Plant introduction
2. Mass and progeny selection
3. Backcross method
4. Heterosis breeding
5. Synthetic breeding
6. Composite breeding
7. Polyploidy breeding
8. Distant hybridisation
9. Transgenic breeding

3. Asexually Propagated

Asexually reproduced plants are propagated by other methods, such as vegetative plant components like stem or root cuttings. These species occur in both self-pollinated and cross-pollinated populations. Most creatures that reproduce asexually are highly heterozygous, have a diverse genetic background, and are extremely adaptable.

1. Plant introduction
2. Clonal selection
3. Heterosis breeding
4. Mutation breeding
5. Polyploidy breeding
6. Distant hybridisation
7. Transgenic breeding

Major Breeding Techniques Used to Improve the Quality of Horticultural Crops

The scientific foundation for genetics and plant breeding was laid by Darwin and Mendel's fundamental discoveries. The basis of contemporary plant breeding, an interdisciplinary discipline, has been enhanced by the recent introduction of developments in biotechnology, genomic research, and molecular marker applications. Enhancing the quality of horticulture crops may benefit society in a number of ways. Elite cultivars may be given desirable genes using this technique, greatly increasing their worth. It has special chances to raise nutritional quality and provide additional health advantages. Many crops have undergone selective breeding to enhance features including improved flavour, increased nutritional value, and reduced bitterness or anti-nutritional effects. Transgenic horticulture can potentially be utilised to distribute vaccines. In this section, many breeding techniques have been discussed in relation to particular successes made in several significant crops (Debnath and Guha, 2015).

1. Plant Introduction

The first technique used in plant breeding is introduction. It is the moving of existing plants (crops) to a newer place where they had never before been produced for human use. The introduction of plants may be categorised based on factors including adaption and use.

Primary: Once an introduced variety has acclimated to the new environment and has not undergone any genotypic alterations, it is immediately released for commercial use in the new location.

Secondary: The best members of the newly introduced variety are chosen by genetic modification. Since individuals have crossed borders, several crop plants have been brought into our nation. For instance, Central Asia introduced the apple, pear, and walnut to various regions of India. Invading Muslims brought grapes and cherries from Afghanistan around the year 1300. The Portuguese also brought potato, chilly, pineapple, and papaya to this country. Cauliflower and cabbage, two well-known vegetables, were imported from the Mediterranean region. Plant introduction is being done in an organised manner under the direction of the Indian government. The NBPGR in New Delhi will play the function as the primary institution. This organisation deals with horticultural and agricultural crop introductions to our nation. Similar responsibilities are carried out internationally by the Rome-based International Plant Genetic Resources Institute (IPGRI).

a. As Variety, Direct Release

The pea cultivar Bonneville, which grows around 120 cm tall, was imported from the United States. The green pods will be ready for plucking in around 90 days and the seeds are wrinkled. The yield per hectare is about 100 q of green pods.

Early Badger (pea) is a cultivar that was imported from the USA and has both early and dwarf varieties. Approximately 7 cm tall and green in colour, the pods are suitable for plucking 60 to 65 days after seeding. The yield is 80–85 Q/ha on average.

Marglobe (tomato): Plants of this late-maturing variety are ambiguous in nature. Large, rounded, smooth, and juicy describe fruits.

Sioux is a high-yielding tomato cultivar from the United States that has a dwarf, spreading, and determinate growing habit. When they ripen, the medium to large-sized fruits have a yellowish green colour. It may be grown in India's mountainous areas. It produces 250 Q/ha, roughly.

b. Choice from the Introduced Material

Japanese white (radish): The flesh is crisp, firm, and delicately flavoured, with a snow-white skin. In 45 days it becomes mature. On plains, it may be grown between October and December, while on hills, it can be grown between July and September.

c. Varieties through Hybridisation

Tomato variety Pusa Ruby: This variety resulted from a Meeruti and Sioux cross. The plants are determinate and medium-sized. The production is around 300 Q/ha, and the fruits are medium in size, lobed, and consistently crimson when mature.

The tomato cultivar Pusa Early Dwarf resulted from a hybrid between Red Cloud and Meeruti. An early dwarf cultivar that was imported from the USA is called Red Cloud. Fruits develop 60 to 70 days after transplanting and are medium in size, evenly crimson, and ribbed. There is a 300-400 Q/ha yield.

Carrot PusaKesar: PusaKesar is a variety derived from the cross between Local Red and Nantes Half Long. PusaKesar has a high carotene content of around 38 mg/100 g. A remarkable quality of this variety is that the roots stay in the field for roughly a month longer than Local Red without exhibiting any signs of bolting. It can withstand temperatures that are greater than Nantes.

2. Pureline Selection

The offspring of a single, homozygous self-pollinated plant are referred to as pure lines. With the help of the single best (elite most) plant progeny among the established varieties or land races, a new variety of plant is created using the pure line selection technique of plant breeding. Pure line varieties are homozygous and homogenous because they are real breeding types with same genetic makeup. These types have a limited genetic foundation, making them less adaptable and more prone to illness. For self-pollinated crops, a pure line breeding technique is typically used. However, it may also be employed to create inbred lines that are then utilised to cross-pollinate or self-pollinate crops to create hybrids (Janick, 2004).

Years of Practice	*Description*
First year	Depending on their genotype, a number of individual plants (between 200 and 3,000) are chosen from a local variety or another mixed population. Always harvest seeds individually
Second year	Each seed from a single plant is cultivated individually in different rows with the appropriate spacing. Here, the lines are assessed for the characters being taken into account. Although it may not always be able to evaluate each line individually in this first season, aesthetically unattractive, weak, and deformed progeny are definitely discarded. In this phase, the number of progenies is substantially decreased, and fewer are transported to the following step.

Years of Practice	Description
Third year	A preliminary yield study is conducted using seeds that were gathered in the second year. Unwanted progenies are discarded, whereas standard plots are maintained for comparison.
Fourth to seventh years	Multiple sites are used for replicated yield experiments. Unwanted progenies are discarded, whereas standard plots are maintained for comparison.
Eighth year	Best offspring are made available as a new pure line variety. Multiplying seeds for dispersion

Significant Progress in Quality Improvement Areas

Tomato variety ArkaSaurabh: IIHR, Bangalore, created it via selection. Plants grow in a semi-determinate manner. The fruits have a nipple-tipped, firm, medium-sized, deep crimson appearance. Fruits are good for processing as well as for consumption when they are still fresh from the market.

It is a variety of the mixed Batia cultivar cultivated in Punjab, Delhi, and western Uttar Pradesh in India. Pusa Purple Long (brinjal) It is a cultivar with an early maturation period, with ripe fruit suitable for harvesting in 100–110 days. Fruits are shiny, purple, and soft. They are long (20–25 cm). The yield is 281 Q/ha on average.

Pant T-3 (tomato): Pantnagar Agricultural University used pure line selection to create this variety. The plants have big, semi-determinate leaves. It is suited for wintertime sowing. Pant T-3 fruit is excellent for processing. The yield is 200 Q/ha on average.

3. Pedigree Selection

It is among the most widely used breeding techniques globally. This approach is named in such a way because its variations call for extensive documentation, *i.e.*, the keeping of meticulous records on the ancestry of the chosen people or lines. These data let us know which F 2 plant the chosen progeny came from. To build near pure lines using this technique, fundamentally a plant to row system is also modified, then the lines' performance is tested.

Steps Involved	Description
Hybridisation	Hybridization between parents who are better
F 1 generation	F1 seeds, which are obtained by hybridization, are sown with the correct sowing distance. The bulk harvest of seeds from 20–40 plants is forwarded to the F 2 generation.
F 2 generation	The primary tool used in this stage is selection. From F1 seeds, around 10,000–20,000 plants are produced (F 2 plants). Separate selection and harvesting of 500 plants approximately.

Steps Involved	Description
F 3 generation	Each F2 generation plant is nurtured to produce about 30 to 40 offspring. They choose between 100 and 400 outstanding plants, albeit fewer are preferred than those chosen in the F 2 generation.
F 4 generation	Spaces are sown with seeds from the F3 generation. Plants with favourable traits are chosen far less frequently than those from the F3 generation.
F 5 generation	Plots with many rows (3 or more) are used to sow individual plant progeny so that superior plants (about 50–100) can be chosen through comparison.
F 6 generation	Plots with three or more rows are planted with individual plant progeny. Plants are chosen based on eye inspection; offspring that exhibit segregation should be removed.
F 7 generation	Preliminary yield tests with a check and a minimum of three replications. There are tests for quality.
F 8 to F 12 generation	Replication-based multi-location yield studies are carried out. Testing is done for disease resistance and product quality.
F 10 or F 13 generation	Increased seed production for distribution

Important Achievements

PusaKranti is a lesser-seeded cultivar of brinjal with a dwarf and spreading growing behaviour. Oblong, dark purple, 15-20 cm long fruits have a bright green calyx. Under the circumstances of North India, the cultivar is suitable for planting in both the fall and spring.

The tomato cultivar Pusa Early Dwarf was created by pedigree selection from a Meeruti and Red Cloud hybrid. With a height of 50–55 cm, this cultivar is dwarf in nature. Fruits are ribbed, medium in size, evenly crimson, and flattened. After transplantation, fruits are ready to be picked 60 to 70 days later. There is a 300-400 Q/ha yield.

4. Heterosis Breeding

Heterosis is the superiority of F 1 above its better-parental, mid-parental, and standard varietal values in one or more traits. The creation of hybrid varieties for the genetic enhancement of a character that needs improving is known as heterosis breeding. In hybrids, heterosis is fully used; in synthetics and composites, it is only partially utilised. Commercial hybrid seed production necessitates a number of crossing methods that are tailored to the floral biology and blooming behaviour of the crop. For the purpose of producing commercial hybrid seeds, a few crossing processes are used.

Hand Emasculation and Pollination

Hybrid seeds are manually created. It is common practise to remove the male organ from the female plant before the commencement of anthesis in order to alter the flower structure. However, this approach is only appropriate when a single flower or plant has distinct male and female reproductive organs. Bisexual beautiful blooms are practised with this technique. The anther (the male portion) is removed from the female line to make it sterile, and the line is then employed as female line to be pollinated by the chosen male parent source.

Self-Incompatibility (SI)

By recognising the presence of self-pollen in or on the stigma of the female pistil, self-incompatibility defeats self-fertilization. When pollen from other plants is delivered by various physical agents, such as the wind, pollinators, *etc.*, successful pollination and fertilisation take place. Therefore, self-incompatibility encourages cross-pollination while preventing self-pollination (inbreeding) (out breeding). One or more multi-allelic loci regulate it genetically, and a sequence of intricate cellular biochemical interactions between the pistil and self-incompatible pollen carry it out.

Dicliny

In unisexual flowers, there are two forms of diplonia. Male and female flowers are produced on the same plant in monoecious plants, such as cucurbits, maize, castor, and coconut. Male flowers in dioecious plants are produced on several plants, such as papaya and cannabis.

Male Sterility

The fundamental prerequisite for hybrid breeding is a female plant in which no viable male gametes are borne. The goal of emasculation is to turn a plant into a pure female by removing all male reproductive components, such as pollen grains. However, doing manual emasculation for large-scale commercial F 1 seed generation is not very simple. Finding or creating a line that is incapable of producing viable pollen serves as a replacement. Because the biological necessity of an external pollen source depends on inhibiting self-pollination in female plants, this method is a trustworthy one. The inability of plants to generate functioning anthers, pollen, or male gametes is referred to as male sterility. Male sterility may be cytoplasmic, genetic, or both types of cytoplasmic. Cross-pollination is encouraged and self-pollination is prevented. Nuclear or cytoplasmic genes, or both, can regulate sterility. In the generation of commercial hybrid seeds, female plants are crossed with male fertility restorer line (R) and male-sterile line (A) to create F 1 hybrids (Rodomiro, 2012).

Important Achievements

A hybrid cultivar of brinjal called Pusa Anmol was created by crossing Pusa Purple Long and Hyderpur at the IARI in New Delhi.

Tomato-72 Sweet It is a hybrid of "Sioux" and "Pusa Red Plum." Fruits are spherical, consistently crimson, less acidic, and have green stem ends. The variety may be grown in Central India and is semi-determinate.

PG-13 (pea): This hybrid pea variety is appropriate for cultivation in Punjab. It is a small plant with white blooms and creamy-white, somewhat wrinkly seeds. It takes 125 days to reach maturity and produces 25 Q of dry grains per hectare.

The Trivandrum-based Central Tuber Crops Research Institute (CTCRI) created the sweet potato hybrid H-41. This fiber-free cultivar was created by crossing a local kind with the Japanese variety Norin. The meat boils readily and has a delicious flavour. The crop may be harvested in 120 days, with a potential yield of 370 Q/ha of tubers.

H.42 (sweet potato): The CTCRI in Trivandrum created this hybrid as well. It is a fiber-free cultivar as well. The hybrid was created by crossing the native cultivar Vella Damph with the American variety Triumph. The hybrid has the ability to withstand weevil assault. It produces 350 Q of tubers per hectare.

Kohir Safeda (guava) is a hybrid produced by crossing Kohir with Allahabad Safeda. It has a few soft seeds and white flesh. Larger fruits are produced by the variety's robust tree-like growth.

Hisar Safeda (guava): This heterotic hybrid, which has upright growth and a compact crown, is a cross between "Allahabad Safeda" and "Seedless." Its spherical, 92-gram fruits have a creamy white flesh with few hard seeds, 13.4 per cent TSS, and 185 mg of ascorbic acid per 100 grammes.

The guava variety Hisar Surkha is a cross between Apple Colour and Banarasi Surkha. The crown of the tree is wide to compact and medium in height. The spherical fruits weigh around 86 g apiece. The pulp is pink in colour and contains 169 mg/100 g of ascorbic acid in addition to 13.6 per cent TSS and 0.48 per cent acidity. 94 kg of yield per tree each year.

5. Mutation Breeding

The main prerequisite for plant breeding is genetic diversity. Sometimes the attribute being improved does not experience the desired variation. Successful use of mutagenic agents (radiation or chemicals) can result in genetic variances from the current population that can be used to select desirable mutants. It has been demonstrated that induced mutation may be used to create diversity within a crop variety (Singh, 2005).

Physical Mutagens

Type of Radiation	*Properties*
X–rays	Sparsely ionising (SI), penetrating and non-particulate
Gamma rays	SI, very penetrating and non-particulate
Alpha particles	Densely ionising (DI), particulate, less penetrating and positively charged
Beta rays particles	SI, particulate, more penetrating than alpha particles and negatively charged

Type of Radiation	Properties
Fast and thermalneutrons	DI, particulate, neutral particles, highly penetrating
Ultraviolet rays	Nonionising, low penetrating

A Description of the Mutant Breeding Process

Year	Generation	Characterisation
1st		Mutagens are either physical (radiation) or chemical agents that are used to treat seeds, pollen, vegetative portions, or tissue cultures.
2nd	M 1 (M 1 V 1)	Plants produced from vegetative propagules or treated seeds (M 1) (M 1 V 1)
3rd	M 2 (M 1 V 2)	Plant population cultivated from vegetative parts (M 1 V 1) or seeds (M 2) obtained from M 1 or M 1 V 1, respectively. Desired mutants may be chosen in this generation or one that follows.
4th–9th	M 3 –M 8 (M 1 V 3 –M 1 V 8)	Continued selection, genetic verification, multiplication, and stabilisation of mutant lines' field performance
10th–13th	Next 2–3generations	Comparison studies of mutant lines conducted throughout time and in various places
14th onwards	Next 2–3generations	Before being released as a new variety, official testing

X-Rays

Rontgen made the discovery of X-rays in 1895. They are extremely penetrating and only weakly ionising. They are produced by X-ray equipment. Chromosomes can be broken by X-rays, which can also cause addition, deletion, inversion, transposition, transitions, and transversions—all forms of mutations at the nucleotide level. Muller employed X-rays for the first time to induce mutations in Drosophila in 1927. In order to induce mutations in barley, Stadler utilised X-rays for the first time in plants in 1928.

Gamma Rays

Compared to X-rays, gamma rays have a shorter wavelength. It penetrates further than X-rays. They are produced by the radioactive decay of several elements, including 14C, 60Co, radium, and others. Cobalt 60 is one of them that is frequently used to produce gamma rays. Chromosome and gene mutations are also brought on by gamma radiation.

Important Achievements

The NBRI, Lucknow, is the first organisation to study somatic mutation in chrysanthemums. They do this by using a Cobalt 60 (60Co) radioactive gamma irradiation source to breed a number of lines by changing one or few characters of an exceptional cultivar without changing the remaining and/or unique part of the existing genotype. In order to create new types of roses, mutation breeding has also been utilised effectively (Wilde *et al.*, 2012).

6. Clonal Selection

Clones are plants created by vegetative propagation from a single plant. The process of selecting the desired (better) clones from a batch of clones for cultivation is known as clonal selection. Cuttings from roses, potato tubers, onion bulbs, and banana suckers are examples of clones.

1. Clones have identical genotypes and comparable phenotypes.
2. Genotypical clones are heterozygous, allowing for the long-term exploitation of hybrid vigour.
3. For whatever number of generations, clones maintain their genetic stability.

After conducting field experiments, healthy plant sections from healthy plants were chosen based on desirable phenotypical traits. Except when there are desired mutations, selection inside a clone is less successful. The clones are always chosen at random. The finest performing superior clones are put through three years of testing at many sites before being introduced as new kinds (Van and Gardiner, 2014).

Mango: After a thorough investigation of the Dashehari orchards in the area of Maliabad, Uttar Pradesh, the best clone, Dashehari-51, with its better yield and regular bearing potential, was isolated.

7. Biotechnological Methods

Plant biotechnology refers to a range of interconnected technologies that evolved from traditional plant selection through a phase in which humans manipulated the cellular environment and its biological functions for the purpose of genetic modification and advancement. To increase the quality of horticulture crops, a variety of biotechnological techniques such as recombinant DNA technology, genetic engineering (GE), transgenic modification, or genetic modification (GM) are used. In a nutshell, given the characteristic in question, genes from one species are transferred into other species across boundaries. There are several methods for developing transgenic organisms, including chemical processes, electroporation, particle (microprojectile) bombardment, and transformation mediated by Agrobacterium (Bisognin, 2011).

Chemical Procedures

Polyethylene glycol is used to treat plant protoplasts, and after that, the DNA of their surrounding media is quickly taken up by the protoplasts. The foreign DNA segments can be successfully taken up and stably incorporated into the plant's chromosomal DNA. Protoplasts are then grown in culture under ambient growth conditions, which enabled them to produce shoots and roots, cell walls, and calluses before eventually regenerating into whole plants.

Electroporation

In order to introduce foreign DNA into plants via the electroporation method, protoplast is employed since plant cell walls hinder the transport of macromolecules across them. The DNA of interest is positioned between electrodes in an electroporation cuvette, and protoplast suspension is administered using electrical pulses. Temporary micropores form in cell membranes as a result of electrical pulses, allowing DNA to enter the cell and eventually reach the genome.

Particle Bombardment

To transfer foreign DNA into cells, a method called particle bombardment is utilised. The process uses gold or tungsten particles (1-2 m) that have been coated with the desired DNA. In this scenario, coated particles are loaded into a particle cannon and propelled at high speed (using pressurised helium gas or electrostatic energy from a water droplet exposed to high voltage) toward the target, which might be plant cell suspensions, callus cultures, or tissues. The plant cell walls and membranes serve as entry points for the projectiles. After microprojectiles successfully enter the cells, the surface of the particle releases the desired DNA, which is then incorporated into the genome of the plant.

Agrobacterium-Mediated Plant Transformation

A plant pathogenic bacteria called Agrobacterium tumefaciens is present in soil. A number of dicotyledonous plants are affected by the "crown gall disease" it produces. This bacteria causes plants to grow enormous, tumor-like swellings (galls). Following infection, the bacteria passes a portion of its DNA to the cells of the plant. The effective integration of this bacterial DNA into the plant's genome causes the development of tumours and the corresponding metabolic changes. Thus, it is a genetic engineer by nature. The generation of transgenic plants uses the bacterial transformation principle, and the most popular technique for plant genetic engineering is Agrobacterium-mediated transformation.

Important Achievements

Application of genetic engineering to regulate fruit ripening: Scientists have created a tomato variety that has been genetically manipulated to sluggishly ripen. Numerous fruits contain pectin, which chemically contributes to the hardness of the fruit. An enzyme known as polygalacturonase breaks down the pectin in ripening tomatoes. The tomato fruit's cell walls separate when the pectin is consumed, finally softening. The quantity of this enzyme in tomatoes was lowered by the researchers, which reduced the pace of cell wall deterioration and resulted in a firmer fruit that lasted longer. The FlavrSavr tomato therefore grows for a longer period of time. As a result, sugars are moved to the fruit, boosting its sugar content and adding flavour (Chebet*et al.*, 2002)

Lists of some Outstanding Studies on Improving Quality through Genetic Modification

Target Traits	*Crop*	*Achievement*	*Method applied*
Vitamins and carotenoids	Potato	Significant increases in the amounts of carotenoid, with up to 14 times more -carotene	The beta-epsilon branch of carotenoid biosynthesis's initial stage being turned off
	Sweet potato	Quantitative trait loci (QTL) for dry matter, starch content, and -carotene content identification	
	Lettuce	Increased enzyme activity and tocopherol's conversion to its more effective form. (Vegetable crops' conversion from -tocopherol to -tocopherol might boost their worth and significance for human health because vitamin E lowers the risk of a number of major illnesses)	Improving tocopherol composition by ove rexpressing an Arabidopsis thaliana cDNA encoding -tocopherol methyltransferase
Folates	Tomato	Fruits have almost 25 times more folate than vegetables.	GTP cyclohydrolase I, which catalyses the first step in the production of pteridine, and amino deoxychorismate synthase, which catalyses the first step in the manufacture of p-aminobenzoate (PABA), are ove rexpressed only in fruit
Flavonoids	Tomato	New flavonoids are produced in tomato fruit.	Stilbene synthase, chalcone synthase, chalconereductase, chalconeisomerase, and flavone synthase are examples of heterologous flavonoid pathway genes.
Sweet taste	Tomato	Tomato fruit flavour is sweetened	Using the thaumatin gene from the African plant katemfe to transform tomatoes (*Thaumatococcus daniellii*)
	Potato	Plants that make insulin from potatoes	The expression of the globe artichoke genes 1- SST (sucrose: sucrose 1-fructosyltransferase) and 1- FFT (fructan: fructan 1-fructosyltransferase) results in the production of insulin.
Tearless onion	Onion	Tearless onion	By employing RNA interference to silence the lachrymatory factor synthase gene. It was prevented from converting precious sulphur compounds into the tearing agent.

Future Scope

The aforementioned breeding advances basically started new businesses. They emphasise the idea that the main breeding goal for horticulture crops must be a special quality rather than yields in and of themselves. These characteristics, which are characterised by quantum improvements in quality and consumer-friendly qualities, rapidly increase market demand for these novel cultivars and pave the way for future possibilities in horticulture breeding. The key to horticultural advancement is the innovation of individual breeders paired with further breeding attempts to promote utility and adaption. It is obvious that a method must be developed to include customer happiness in the decision-making process. The sector has to fight marketing initiatives that try to muddle cultivar identification and turn horticulture crops into commodities.

REFERENCES

Bisognin, D. A. (2011). Breeding vegetatively propagated horticultural crops. *Crop Breeding and Applied Biotechnology, 11*, 35-43.

Bliss, F. A. (2010, August). Education of the next generation of plant breeders for horticultural crop improvement. In *XXVIII International Horticultural Congress on Science and Horticulture for People* (*IHC2010*)*: Colloquia and Overview 916* (pp. 181-189).

Chebet, D. K., Okeno, J. A., and Mathenge, P. (2002, August). Biotechnological approaches to improve horticultural crop production. In XXVI International Horticultural Congress: Biotechnology in Horticultural Crop Improvement: Achievements, Opportunities and 625 (pp. 473-477).

Debnath, S., and Guha, S. (2015). Breeding methods for quality improvement in horticultural crops. In *Value addition of horticultural crops: recent trends and future directions* (pp. 201-211). Springer, New Delhi.

Janick, J. (2004, September). Horticultural plant breeding: past accomplishments, future directions. In *International Symposium on Harnessing the Potential of Horticulture in the Asian-Pacific Region 694* (pp. 61-65).

Kumar, N. (2006). *Breeding of horticultural crops: principles and practices*. New India Publishing.

Prohens, J., Nuez, F., and Carena, M. J. (2008). *Handbook of plant breeding*. New York: Springer.

Rodomiro, O. (2012). Transgenic vegetable breeding for nutritional quality and health benefits. *Food and Nutrition Sciences*, 3 (9), 11

Singh, B. D. (2005). Mutations in crop improvement. *Plant Breeding, Principles and Methods. Kalyani Publishers, Ludhiana*, 698-731.

Sonah, H., Deshmukh, R. K., Singh, V. P., Gupta, D. K., Singh, N. K., and Sharma, T. R. (2011). Genomic resources in horticultural crops: status, utility and challenges. *Biotechnology advances*, *29*(2), 199-209.

Van N. S., and Gardiner, S. E. (2014). Breeding better cultivars, faster: applications of new technologies for the rapid deployment of superior horticultural tree crops. Horticulture Research, 1.

Wilde, H. D., Chen, Y., Jiang, P., and Bhattacharya, A. (2012). Targeted mutation breeding of horticultural plants. Emirates Journal of Food and Agriculture, 31-41.

Chapter 9

Participatory Plant Breeding

Rhitisha Sood[1]*, Shailja Sharma[2] and Sanchit Thakur[3]

[1]PhD Scholar, [2]Assistant Professor,

Department of Genetics and Plant Breeding, College of Agriculture, CSKHPKV, Palampur, Himachal Pradesh

[3]Assistant Professor, Shoolini University, Solan, Himachal Pradesh

****e-mail: rhitishasood5529@gmail.com***

ABSTRACT

Developing a plant breeding programme in collaboration with breeders, farmers, marketers, processors, consumers, and policymakers is known as participatory plant breeding. In order to promote on-farm conservation, participatory plant breeding encourages farmers to keep looking for, choosing, and managing regional crop populations that enables them to participate in the creation of new populations/varieties that are more suited to agronomic techniques used in organic farming as well as marginal situations. It outlines the preliminary findings of PPB procedures that handle the challenges of selecting parents and establishing breeding goals across a range of research fields. It is predicated on the notion that professional plant breeders and farmers both possess significant knowledge and abilities that could complement one another. Plant breeders can focus their research on meeting the demands of particular farmer groups (women, men, rich and poor). In the PPB paradigm, the research institution is frequently, but not always, responsible for the creation of genetic variability. The process follows the series of steps i.e. identifying and choosing target environments and users; selecting genetic and parental material as well as breeding methods, naming of varieties; share findings; and disseminate them. Gains in production, improved biodiversity, and cost- and time-effectiveness are all potential outcomes of PPB. Because breeding is done in the target environment, genotype x environment interactions are positively employed in every cross, and at least one parent in every cross is well adapted to the local environment. Due to the adaptation of the local parental materials to the local year-to-

year fluctuations, the effect of the genotype x year interaction is likely lessened. The study also considers the demands of farmers, conservation concerns, and challenges related to diverse users. In order to advance the rights and interests of farmers, this brief aims to increase awareness of participatory plant breeding (PPB) as an extra, complementary, and effective technique.

Introduction

The term PPB first used at a 1995 workshop held by Canada's International Development Research Center, (IDRC). In 2000, PPB became an essential component of each CGIAR center's plant breeding projects. Participatory Research and Gender Analysis (PRGA) is a concept that was first proposed in 1996. PPB, which stands for Participation and Plant Breeding, refers to a "multidimensional, dynamic process of contributing, influencing, sharing, or redistributing power and of control, resources, rewards, information, and skills to be gained through beneficiaries' involvement in decision-making." It is a programme for plant breeding that involves numerous partnerships between breeders and scientists and farmers, marketers, processors, consumers, and policymakers. PPB regards the farmers as partners and gives them a bigger voice in the selection process, enabling them to choose which cultivars are most suitable for them to improve the genetics of a crop. Additionally, PPB has the potential to enhance traditional landraces that would otherwise disappear from the system. In doing so, it gives farmers the knowledge and chances to look for new diversity, as well as to choose and trade populations with characteristics that meet their local tastes and needs. Farmers can participate in PPB in a variety of ways, including by defining breeding objectives and priorities, choosing or providing sources for germplasm, hosting trials on their property, choosing lines for additional crossing, discussing results with the scientists, planning for the upcoming year's activities, suggesting methodological changes, and multiplying and commercialising the seed of the selected lines. On the other hand, the insertion of just a few key genes, such as those relating to plant height or pest resistance, may be what makes PPB products successful. It is also generally believed that by diminishing genotypic and/or allelic variation within the cultivated crop, the use of contemporary cultivars to replace farmers' distinctive varieties may have an impact on the intraspecies genetic diversity. A potent technique that can be applied in high potential production systems is the development of novel varietal variety by PPB, which has already been used effectively in challenging situations.

In conventional plant breeding (CPB), the process is often supply-driven, and new varieties are issued before it is known whether or not the farmers would enjoy them. The delivery phase of participatory plant breeding (PPB) is reversed since the technique is demand-driven and initially adopted by farmers at the conclusion of a full cycle of selection.

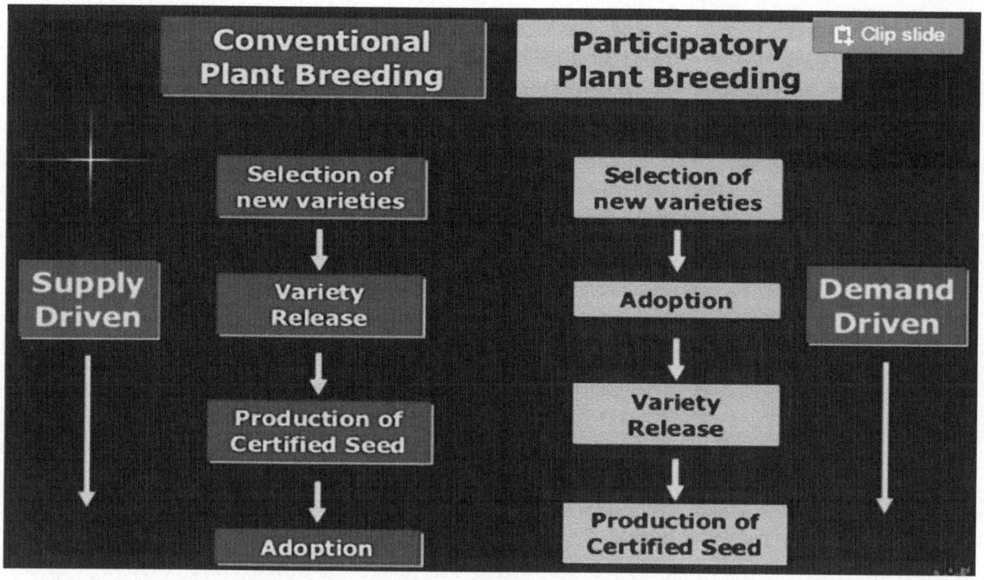

Figure 9.1: A Comparison between Traditional and Participatory Plant Breeding

History

Plant breeding as a practise predates agriculture itself by around 10,000 years, when farmers domesticated crops like barley and emmer wheat. Mendel's early 1900s genetic inheritance experiments are more recently credited with introducing plant breeding as a field of study. A "science-based technology," plant breeding seeks to provide farmers with improved cultivars by way of selection among genetically diverse plant populations. PPB is just one of many approaches that evolved from criticisms of the failure of development initiatives intended to introduce modern agriculture technologies to regions without these resources that started in the 1950s. Apodaca (1952), for instance, describes the USDA extension agent's unsuccessful attempt to get a farming community in New Mexico to switch from a low-yielding traditional maize variety to a high-yielding hybrid that was unpalatable due to its flavour, texture, and appearance. Interdisciplinary approaches were developed in an effort to involve experiments on farmers' fields throughout the research process, encouraging feedback from farmers at various stages of the project. Examples include Farming Systems Research and the Farmer-Back-To-Farmer model. They did, however, stand in stark contrast to the predominant model used at international research centres like the Consultative Group of International Agricultural Research (CGIAR), which Biggs (1990: 1481) referred to as the "central source model." The aim of these centres was to create agricultural innovations that would only reach the farmer after being transmitted to national research systems and then extension agents. The International Potato Center (CIP) in Peru was one of the first institutions to actively promote the notion that farmer knowledge was equally vital to achieving its purpose as official research in the middle of the 1970s (Biggs, 1990). The International Center for Tropical Agriculture (CIAT) in Colombia and the International Rice Research Institute (IRRI) in the Philippines were two such institutions that hosted small teams of researchers who used participatory techniques (Chambers, 1989). Participatory research models were being successfully used by a variety of national agricultural research stations, non-governmental organisations, and farmers' organisations in developing nations by the early 1990s.

Kinds of Participation in PPB

In relation to PPB, alternative terminologies such as "participatory crop improvement (PCI)," "collaborative plant breeding (CPB)," and "farmer participatory breeding (FPB)" are sometimes used interchangeably, where PCI is occasionally divided into two categories.

The first is participatory varietal selection (PVS) refers to procedures in which farmers only choose pre-existing varieties or stabilised (homozygous) materials that are undergoing field testing that they believe are best suitable for their own usage, and the second is participatory plant breeding (PPB), in which breeders and farmers work together to create new varieties for breeding programmes. While selection occurs in PPB within segregating populations, it occurs in PVS among advanced or genetically stable populations and lines. It's not always easy to tell the difference between PVS and PPB, though. Selection among populations in PVS is typically mixed with within-population selection PPB, especially in cases of cross-pollinating populations. Due to the fact that farmers are more actively involved in PPB and make decisions at an earlier and more fundamental stage of the variety creation cycle, PPB has a greater impact on farmer empowerment than PVS. But, before continuing, it is crucial to remember that farmers' interests rarely cease with the assessment of better materials in the outcomes of PPB or PVS.

Functional Participation

Plant breeders can focus their research on meeting the demands of particular farmer groups (women, men, rich and poor). Farmers can reassure plant breeders that their evaluation of the trade-offs between characteristics is accurate. A variety's ability to produce well under "real world" circumstances is guaranteed by on-farm study. The researcher, the farmer, or both may be in charge of managing on-farm research. PPB guarantees that farmers will adopt innovations more successfully. Depending on the crop, parent materials, target location, researcher's ability to assimilate farmer criteria, farmer's ability to handle various sorts of materials, qualities of interest, and size of breeding program/number of materials to be screened, farmer participation can be valuable at various phases. Farmers can play a variety of roles in participatory plant breeding, including providing technical leadership, key social organisational leadership, information sharing, training and skill development, field labour, input supply, and providing landrace or farmer material for additional breeding efforts.

Participatory Variety Selection (PVS)

It is a procedure whereby the partners participate in the field testing of finished or nearly finished varieties, often a small number. Any breeding program's main activity, selection, takes place throughout the whole breeding process. Consequently, PVS is a constant component of PPB. Participatory varietal selection can be used to successfully identify variations that farmers will accept, removing the barriers that force farmers to grow outmoded or outdated types. PVS can also be used as an initial step, sort of an exploratory test, to assist partners in accurately estimating the level of time and land commitment that a fully developed PPB plan requires.

Witcombe *et al.* (1996) describe both of these methods for the first time in the peer-reviewed literature, specifically referring to PPB as "a logical extension of participatory varietal selection," in which farmers are involved in the earliest stages of selection from segregating populations. With a growing number of successful participatory projects, the CGIAR began to recognize the value of participatory research and formalized its commitment to this process with a systems-wide initiative on Participatory Research and Gender Analysis (PRGA) in 1996. By 2000, a recommendation made to the CGIAR Technical Advisory Committee suggested "that PPB become an integral part of each CGIAR center's plant breeding program".

	F	S	F	S	F	S	F	S	F	S	F	S
Selection of Source Germplasm	✓		✓	✓	✓	✓	✓	✓		✓		✓
Trait Development (Pre-breeding)	✓		✓	✓	✓	✓		✓		✓		✓
Cultivar Development	✓		✓		✓	✓		✓		✓		✓
Varietal Evaluation	✓		✓		✓	✓	✓	✓	✓	✓		✓
	Model 1: Traditional Farmer Breeding		Model 2: Grass-roots Breeding		Model 3: Complete Participatory Breeding		Model 4: Efficient Participatory Breeding		Model 5: Participatory Varietal Selection		Model 6: "Conventional" Plant Breeding	
Note: F=Farmer; S=Scientist	FB		GB		PPB		COB		PVS		CPB	

Figure 9.2: Different Approaches to PPB based upon Participation in Breeding Process (Halewood *et al.*, 2007).

How does PPB Contribute to Farmers' Rights?

According to Article 9.2 of the Treaty, Contracting Parties should support farmers' rights by safeguarding traditional knowledge about plant genetic resources for food and agriculture (PGRFA), compensating farmers for using PGRFA, and allowing farmers to participate in national discussions about PGRFA conservation and use. PPB supports farmers' rights in a number of ways by giving them chance to have a say in how technologies are developed in ways that take into account their unique demands, agro-ecological surroundings, and cultural preferences. Commercial conventional breeding frequently focuses on creating cultivars with high potential for production that are suited to a wide range of non-extreme situations for resource-rich farmers but these may not be suited to extreme environments tagged along with the marginal farmers. Non-commercial conventional breeders, like state-owned extension stations and breeding/research institutions, frequently lack the resources required to meet the needs of resource-poor farmers, including access to environments similar to those found on farmers' fields and a lack of understanding of the particular needs of farmers. As a result, they might not be able to provide for farmers who produce under difficult conditions. The participatory approach is evident in both the location of the (field) research as well as the varied backgrounds of the individuals involved in the selection and breeding process. Local communities have test fields, allowing novel types to adjust to actual production circumstances. Another strategy to maximise adaptation to the local agro-ecological conditions is to choose the genetic material that is utilised in collaboration with farmers and has frequently already been grown in the community. PPB also entails giving farmers access to genetic variety so they can use it as a foundation for innovation. PPB can improve farmer seed systems, which are the methods by which farmers produce, pick, save, and obtain seeds. Four key traits define a healthy seed system: I it maintains a base of germplasm that offers diversity, adaptability, and a basis for selection; (ii) it produces quality seed for production (free of disease, with high germination rates, and vigorous); (iii) it ensures seed availability and distribution (seed sources, social networks, markets); and (iv) knowledge and information about the seed are available and shared (growing methods, utilization, knowledge of new materials).

Additionally, as the participative process advances, the farmers experience a growing sense of ownership as their chosen breeding lines are used to create crosses and begin a new breeding cycle. A cascading cascade of local breakthroughs can result from such empowerment. Though PPB is best suited for creating goods for use in harsher situations, it is also effective at creating materials for use in settings with great potential for manufacturing yield. PPB makes advantage of the farmers' accumulated traditional knowledge, which it raises the prominence of the information and they have a chance in decision making for money allocation in terms of agricultural research and extension activities. Farmers gradually improve the quality of their engagement as a result of this empowerment impact, turning them into actual research partners. As a result, PPB promotes capacity building and knowledge and technology transfer for both the farmers and the formal-sector breeders they interact with. Farmers are involved not only in breeding activities but also in the registration of the variety produced, its maintenance, seed multiplication and distribution, and, as necessary, commercialization in more broadly conceived PPB projects that focus on downstream use and dissemination of PPB products. In some circumstances, it would be important to look into gaining formal rights over such documents so that the involved farmers, their associations, or their communities can be recognised for their efforts through the participatory process. Legal restrictions and standard-setting organisations have a direct impact on farmers' capacity to certify, multiply, and distribute seed in many nations. Therefore, PPB and PVS can and should educate farmers about these regulatory frameworks if they prevent them from maximising the benefits of utilising the materials they helped to improve. They should also, whenever possible, involve farmers in efforts to influence the modification of these frameworks. By doing this, PPB greatly increases the possibility that the end products will be suitable for the circumstances of the involved farmers and hence better contribute to their lives.

Goals of PPB

- Boost crop yield to make it more profitable with direct benefits to the user
- Enhance farmer selection and seed production efforts by improving farmer capabilities.
- Advancing techniques for biodiversity and germplasm conservation
- Creating adapted genetic material, especially for underserved consumer groups (*e.g.*, women, poor farmers)
- Decentralize programmes to increase the cost-effectiveness of breeding programmes and their ability to target new niches, and also allowing various participants in a breeding programme at various stages.

Importance of PPB

In traditional plant breeding, breeders focus on classical breeding from breeding goals like prioritised traits, germplasm selection, breeding techniques, and performance evaluation. Despite the dramatic increase in production, conventional plant breeding has a number of drawbacks and hasn't done much to enhance the standard of living for subsistence farmers worldwide. Conventional plant breeding has some drawbacks, including a disconnect from the real world, a small number of beneficiaries, a limited role for real stakeholders, the persistence of hunger and poverty, nutrient availability problems, environmental concerns, and a higher incidence of diseases, that can be overcome through PPB.

Benefits and Limitations of PPB

1. **Increasing adoption rate:** Farmers' collaboration in material identification and selection is likely to result in higher adoption rates. PPB makes advantage of the farmers' accumulated traditional knowledge. PPB can result in crops that not only perform well in their intended habitats but are also available more rapidly and are more readily adopted by consumers by allowing farmers to choose varieties suitable for their particular soil quality, rainfall circumstances, and resource requirements. In comparison to conventional plant breeding, PPB reportedly cut the period for varietal creation and distribution from 10-15 years to 5-7 years.
2. **Improving seed systems:** PPB can help farmers' seed systems, which are their methods for producing, choosing, preserving, and acquiring seeds. Because it allows farmers better access to high-quality seed of their choosing, encouraging and supporting on-farm seed production by farmers is viewed as one strategy for sustainable seed supply in Africa.
3. **Increasing farmers' ownership, empowerment, and rights:** PPB often requires a larger and more intricate level of farmer involvement. PPB can greatly empower participating farmers when they are actively involved in the early stages of research design. When determining community requirements and preferences, it is easier to take into account suggestions from a wide range of community members, not only those of the farmers chosen to participate in the breeding programme, if the varietal selection is done on-farm. In turn, the procedure leads to more farmer and community ownership over the enhanced lines and has the potential to give disenfranchised community members more influence. PPB can be inclusive of women in particular and, more crucially, take into account their preferences and tastes, which frequently differ from those of their male counterparts.
4. **Enhanced genetic diversity:** PPB encourages the creation and upkeep of more genetically diverse cultivars. The PPB-based breeding approach promotes the maintenance of more diversified, regionally suited crops, in contrast to the existing global breeding paradigm, which has primarily focused on creating a small number of varieties that are stable through time and adapted to a wide range of settings. Large financial investments may be necessary to scale up PPB techniques for use at the regional, governmental, or worldwide levels.
5. **Restrictive regulator policy:** The introduction of PPB products into currently operating informal seed systems has the potential to be a very effective and low-cost strategy. However, regulations and policies have a big influence on how much farmers can participate fully in

PPB. Therefore, PPB and PVS can and should educate farmers about these legal frameworks and, where appropriate, encourage their participation in initiatives to have an impact on their adjustment.

6. **Benefited costs relationship:** Due to potential savings on field testing sites, fewer overhead expenses, and a shorter research period needed to produce viable materials, PPB may also be less expensive to carry out than traditional breeding. On-farm crop variety evaluations demonstrated a cost of US$0.50 per recorded data point for participatory trials, compared with $0.80 for conventional trials, according to one study. The benefit/cost ratio of participatory plant breeding (PPB) is 2.6 times higher than that of conventional plant breeding, according to a recent cost-benefit analysis in Syria. But, as farmers participation increases, they are required to devote more time, effort, and money and also to contribute more thoughts and complex analytical techniques.

Both PVS and PPB aim to alter the structure of the local crop population to increase its productivity and usefulness to farmers, but both techniques may have detrimental effects on landrace diversity. Successful PPB products, it has been proposed, may smother a sizable portion of local diversity, boosting production in the short term but displacing unimproved populations in the area, increasing susceptibility in the future. Despite PPB's benefits, acceptance rates for regular or long-term use in national and international breeding programmes have been modest. PPB is best suited to cereals and less-so to roots, tubers, and pulses; the general attitude among scientists, professionals, and decision-makers is that PPB is opposed to conventional plant breeding programmes rather than complementary; and strict seed registration requirements in many countries can be a deterrent to both public and private breeding programmes that ultimately seek to improve crops. PPB must become a constant component of plant breeding programmes, addressing crops cultivated in agriculturally tough and climatically hard settings, in order to be a sustainable strategy to plant breeding. Furthermore, before a variety can be recognised in a rising number of nations, it must meet relatively high standards for distinctness, stability, and notably uniformity. Even though they may be very beneficial to farmers, some PPB products won't be able to meet these requirements.

Activities included in PPB

PPB can involve a variety of tasks, some of which are : (i) Identification of breeding goals (ii) Generating genetic diversity (includes the provisioning of genotypes to be included in breeding program) (iii) Experimental varieties are developed by selection among distinct populations (iv) evaluation of the experimental varieties (v) release of the variety (vi) farmer promotion of the newly developed variety (vii) Production of seeds: The ability of farmers to work on bettering their own circumstances is acknowledged as a key requirement

Mode of participation	Role of plant breeders	Role of farmers	Comments
Farmers are given finished varieties developed by plant breeders	• Set breeding objectives • Select source germplasm • Identify traits for improvement • Determine breeding methodology • Establish testing procedures • Evaluate finished cultivars on station	• Decide only whether or not to adopt the product	• Traditional breeding • Little direct interaction between farmers and breeders • Breeders knowledge of what farmers' want is not based on organized and direct interaction with farmers
Farmers provide source germplasm on which breeding process is based	• Collect and characterize source germplasm • Identify traits for improvement • Determine breeding methodology • Establish testing procedures • Evaluate finished cultivars on station • Basis for developing new varieties	• Provide source germplasm • Part of target population	• Source germplasm comes from farmers in target population, rather than gene bank • Well adapted material, hopefully with many traits farmers value • Tenuous relationship between farmers and breeders • Breeding process solely in the hands of breeders
Farmers identify traits to be improved and suggest selection criteria	• Set breeding objectives • Select source germplasm • Determine breeding methodology • Establish testing procedures • Evaluate finished cultivars on station	• Identify traits for improvement	• Better targeted varieties • Varieties more likely to respond to farmers' needs and constraints
Farmers evaluate finished varieties on station or in scientists-managed on-farm trials and help select varieties to distribute	• Set breeding objectives • Select source germplasm • Identify traits for improvement • Determine breeding methodology • Establish testing procedures • Finished cultivars evaluated on station or in farmers' fields but under breeders' management	• Actively participate in testing procedures • Identify finished or near finished varieties that are interesting to them	• Farmers may be able to select for traits that they cannot easily describe in words • Decision-making and responsibility for selection of germplasm shared between breeders and farmers • If varieties are planted on-farm in several different locations, breeders and farmers can evaluate them under a range of biophysical conditions
Farmers evaluate unfinished materials (lines, families, landraces) on station or in scientists-managed on-farm trials and select materials for further improvement	• Help set breeding objectives • May select source germplasm • Help identify traits for improvement • Determine breeding methodology • Establish testing procedures • Finished cultivars evaluated on station or in farmers' fields	• Identify interesting materials that still show a high degree of genetic variability for further improvement • Help to set the breeding objectives • Identify traits for improvement	• May lead to more diverse set of materials to be improved • Provides good idea of genotype-by-environment interactions if done in farmers' fields
Farmers conduct germplasm evaluation trials in their own fields and using their own management practices	• Help set breeding objectives • May select source germplasm • Help identify which traits will be targeted for improvement • Determine breeding methodology	• Farmers actively participate in testing procedures • Testing is done in farmers' fields and under their management • Identify near or finished varieties or interesting materials that still show a high degree of genetic variability for further improvement • Help to set breeding objectives • Identify traits for improvement	• Materials could be finished varieties or unfinished materials in different stages of improvement • Provides a very good idea of genotype-by-environment interactions • Explicitly incorporates farmers' needs, interests, and constraints • Strong organized interaction between breeders and farmers • Sharing of decision-making, responsibilities, and activities
Farmers are trained in 'scientific' breeding methods	• Train farmers in scientific breeding methods so they can: (1) maintain valued traits in their varieties, (2) modify existing traits, and/or (3) introduce new traits.	• Set breeding objectives • Select source germplasm • Identify traits for improvement • Determine breeding methodology • Establish testing procedures • Testing done on farmers' fields	• Trained farmers are able to carry out breeding process on their own, possibly with help from scientists

Figure 9.3: Modes of Participation in PPB (Morris and Bellon 2004).

for sustainable agricultural development. In order to identify effective PCI approaches and opportunities, it is important to consider various forms of participation and the need to relate these to variations in the institutional context (farmer-led, formal-led PCI) and bio-socio environmental context (variation in agro-ecological and economic conditions). Farmers can take part in plant breeding at various stages by serving in a variety of leadership roles, including technical leadership, key social organisational leadership, information sharing, trainer/skill building, field labourer, input supply, and providing landrace or farmer material used for further breeding work.

Components

Choice and Selection of Parents

Understanding the distribution of variety according to farmers' use values in the selection of parents (local landrace) is one of the crucial first steps in the PPB project. In a breeding programme, the best breeding material chosen from the previous breeding cycle makes up a large portion of the parental material used in crosses, and because in PPB the selection is made by both breeders and farmers, farmers do in fact take part in the selection of the parents to start a new breeding cycle. The conflict of interest between selecting parents and establishing breeding objectives is brought to light through a consultative participation process used to evaluate farmers' requirements and the project purpose. To prevent bias from powerful organisations in the society, the parents were chosen based on the priorities and interests in the community, goals and objectives of the programme, technical viability, genetic variability availability, and consumer interest, focus group discussions (FGDs) with the aim of determining the landrace parent for PPB. Participatory rural appraisals (PRAs) were conducted to (1) understand major variables influencing farmers' decisions to keep local cultivars, including information on market development, and (2) identify key impediments to production systems relying on local crop cultivars. Before beginning to construct a PPB programme to encourage on-farm conservation, plant breeders must fully comprehend such distribution patterns and the underlying decisions made by farmers. Farmers and other stakeholders participate in various stages of selection in their fields, and there is ongoing communication between the research institute and other farmers taking part in the PPB initiative. In each place, the selection is carried out independently. This typically results in the selection of unique entries at unique places, although it does not preclude the selection of the same material. Farmers can employ the best breeding stock created after the four phases of selection as varieties, and the research centre can use it as parental stock for crosses to start a new breeding cycle.

The Stepwise Process Involved in PPB Modelling

PPB relies on three key elements: (1) a locally adapted cultivar serves as one parent; (2) selection is decentralised in the target environment; and (3) farmers are involved in the plant breeding process.

The steps include:

- ✫ Identify agroecosystems and locate communities that are interested.
  - ❖ Listing of existing local diversity, categorizing the variety of local crops based on the planted area of each cultivar and the proportion of households that cultivate each cultivar
- ✫ Hold "diversity fairs" to identify crops diversity and gather local expertise and germplasm.
- ✫ Utilize a community biodiversity registry to comprehend and track the diversity of local crop types.
- ✫ Conduct preference ranking and paired matrix ranking to conduct assessment analysis for desired and undesirable attributes in order to identify traits needing improvement.
  - ❖ Selecting the best landraces utilising preference matrix ranking, where the favoured features of each landrace parent are recorded, and the traits in need of improvement are carefully examined using paired matrix-ranking methods.
  - ❖ The landrace parents preferred by farmers for inclusion in the PPB program are shortlisted.
- ✫ Examine the potential advantages of regional variety and the dangers posed by genetic degradation.
- ✫ Establish breeding objectives for PPB and parent choice with the involvement of a representative community.
- ✫ Make farmer seed systems stronger for quick dispersal.

Identify Target Environments and Users

The representativeness of the main producing locations for a certain crop in terms of climatic variables (temperature, rainfall, elevation), agronomic techniques, soil types, landscape, *etc.*, is the most evident criterion for the selection of the target physical settings which is closely interlinked to target users (whether to work for different socio-economic environments present in the target area, or to privilege the most difficult environments where farmers have fewer opportunities for market access and where most of the agricultural products are used within the farms or within the community, or to work only for the most favorable, high potential, environments possibly market oriented). Once the parameters are decided upon, selecting the actual locations and users calls for the assistance of partners who have a thorough understanding of both the environment and the users that can be assisted by the staff of the extension service or the provincial research stations.

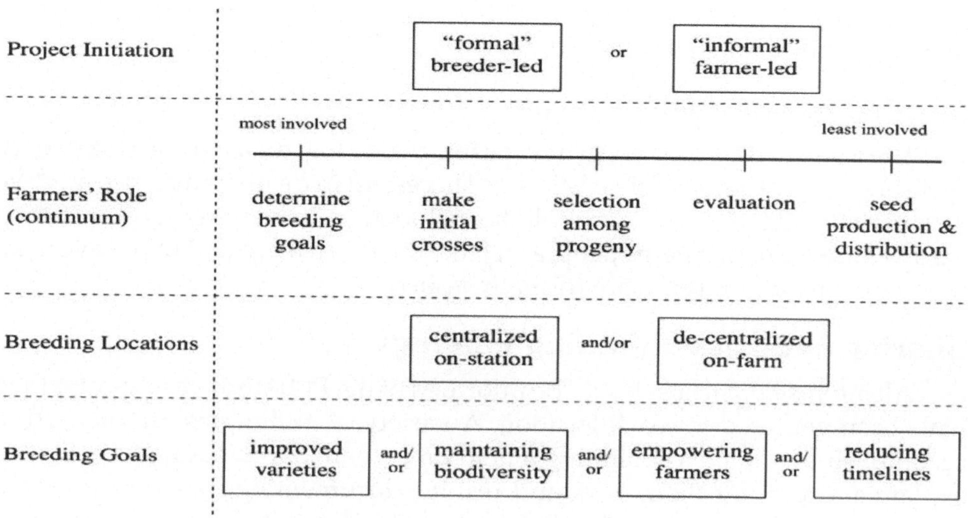

Figure 9.4: Different Components of PPB (Shelton and Tracy 2016).

Choice of Breeding Method

The selection of an appropriate germplasm and the development of objectives are two additional elements that affect the effectiveness of a breeding programme in addition to the breeding method. Without taking into account if and how farmers are managing genetic diversity, the choice of breeding techniques in PPB cannot be determined. Pure lines, blends, hybrids, or open pollinated types are among the breeding techniques that might be used, depending on the desired genetic structure of the finished product. When it comes to self-pollinated plants and pedigree breeding, selection in farmer fields can begin with segregating populations (for instance, F_2-derived F_3 families) after their numbers have been reduced through selection (including MAS) on stations for disease resistance, for traits with high heritability, for quality traits like malting quality, or a combination of these. Field testing can begin as early as the F_3 bulks when the breeding programme employs the bulk-pedigree approach. To gather enough data on the stability and performance of the breeding material for farmers to decide whether to adopt it and for the variety release procedure, the yield testing should be carried out in both situations for at least four successive cropping seasons. Recombination, which can occur on station, while selection and testing might take place in farmer fields, correlates with the formation of genetic variation in the case of population improvement of cross-pollinated crops. Either on station or in farmers' fields, uniform inbred lines can be produced for use as parents of hybrid cultivars. All future generations of vegetatively propagated crops are suitable for testing and selection in farmer fields after the initial crossings, and selecting for traits like disease and/or pest resistance, for traits with high heritability, and for quality attributes can help reduce the number of clones on station.

Naming of Varieties

The issue of how to identify the variety is one that arises at the end of each cycle of the programme if the cycle was successful in developing a prospective new variety. The community should be included in the naming process. When varieties are named, it helps people feel like they own them and may have legal repercussions if PPB types are formally issued.

Sharing and Disseminating Findings

All interested parties should be informed of the PPB study results after they have been gathered for each location. A variety of techniques can be used to accomplish this, such as planning a field day where participating farmers can explain and present their work and results, documenting the work on radio and television, holding stakeholder meetings to discuss the results, instructing participating farmer groups, and creating descriptive sheets for each farmer's chosen variety.

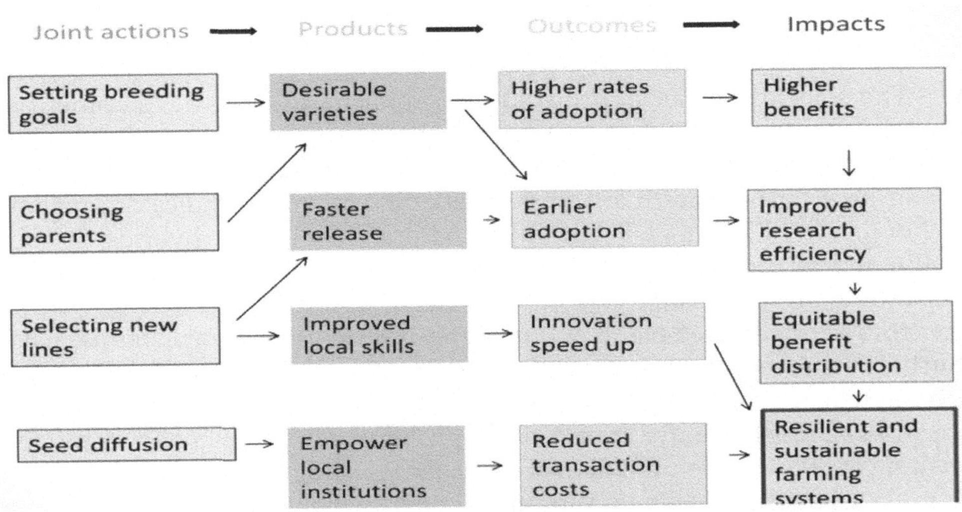

Contribution to Sustainable Agriculture

In developing nations, a large number of small-scale farmers cultivate crops on marginal lands that are either neglected by infrastructure and markets, or have difficult growing conditions, or both. Smallholder farmers are among the populations most susceptible to the negative effects of climate change, especially in developing nations. Improved crop varieties can help to increase food and nutrition security by meeting smallholders' physical and social preferences and being well-suited to a variety of unique growing settings. They can also help to preserve crop diversity. Smallholder farmers can boost their resilience and adapt to climate change by increasing genetic variety and improving crops adapted to arid regions. The empowerment of farmers as well as frequently marginalised community members like women and less affluent farming households is strengthened when farmers are included in decision-making and breeding processes. Participatory plant breeding contributes to sustainable food security and increased resilience to a number of hazards and difficulties by assisting smallholder farmers working in a variety of challenging conditions to develop stronger and more dependable seed systems.

Stakeholders and Organisations Involved in PPB

- Breeders and farmers, marketers, processors, consumers, and policymakers are among the various participants in PPB (food security, health and nutrition, employment). Out of these farmers and breeders work closely together to define goals, make decisions, share responsibilities, and produce the desired results. There are numerous examples in Asia, Africa, and Latin America with national agricultural research organisations and nongovernmental organisations leading additional projects.
- Many governments, non-governmental organisations, NGOs, and research institutions have given this method of breeding the attention it deserves. International Center for Agricultural Research in the Dry Areas (ICARDA), has ongoing PPB programmes in Algeria, Egypt, Eritrea, Iran, Jordan and Syria on barley, wheat, lentil, chickpea and faba bean. International Center for Tropical Agriculture (CIAT), International Maize and Wheat Improvement Center (CIMMYT), National Institute for Agricultural Research (INRA), Local Initiative for Biodiversity, Research and Development (LI-BIRD), Center for Sustainable Development (CENESTA), International Crops Research Institute for the Semi-Arid Tropics (ICRISAT) and the International Rice Research Institute (IRRI) among others, are research organisations that have been involved in PPB projects (Bhargava and Shrivastava, 2019).
- There are currently 25 varieties of barley grown in Syria, where a PPB programme on the crop was launched in 2000. Each of these varieties covers between 5 and 25 000 hectares. In a similar vein, six barley varieties have been identified and bred for their suitability to Egypt's north-western coast, and farmers in Eritrea are brewing three barley and one lentil kinds. Other prosperous nations with various agricultural practises include China (maize), Nepal (rice and maize), Mali (millet), Cuba (maize, beans, rice, cassava, and tomatoes), and Honduras (maize and beans).

- Additionally, it has been shown that PPB goods can be global public goods and are not always of a local nature, as is the case with rice varieties made in Nepal and utilised in India and Bangladesh. Similarly, PPB varieties developed in Syria are also being grown in Iraq. These are easily reproducible and adaptable for usage with a variety of crops everywhere in the world. The state of PPB globally despite the benefits of PPB outlined above, national and international plant breeding programmes have been slow to adopt it over time.
- PPB can be set up to give women participation possibilities even if it is still not a very common practise. Women are playing significant roles in PPB processes in various parts of China where men have moved from farms to towns and cities. Farmers become more aware of the benefits science may provide when they interact with expert breeders through participatory methods.

Discussion and Conclusion

The foundation of participatory plant breeding is the notion that both professional plant breeders and farmers have significant knowledge and abilities that could complement one another. Because it defines the direction of our agricultural destiny, plant breeding has an impact on people and society. Without the proper kinds, farmers cannot be successful, and customers suffer from either rising prices or a lack of food availability, or both, depending on their particular farming methods. Users can participate in all significant stages of the process of plant breeding and variety selection, making PPB in general "participatory." The method for categorising local crop diversity is useful for determining the scope and distribution of farmers' classifications, and it is a straightforward way to introduce farmers to the idea of local common and rare traditional cultivars. PPB achieves this by utilising G x E interaction and choosing kinds specifically for the environment in which they are intended to be used in order to produce improved performance. It is also helpful for examining possibilities for adding further advantages like improved neighbourhood motivation and awareness, seed networks, fresh germplasm and market connections. Social networks demonstrate the efficiency of informal seed systems and decentralised selection. PPB can make good use of nodal farmers' knowledge of genetic material maintenance and selection. Additionally, increasing the nodal farmers' capacity in PPB may significantly increase crop diversification. Participatory techniques like biodiversity fairs, diversity blocks, and community biodiversity registers increase local awareness, enhance local understanding of the importance of local crop diversity, and strengthen the roles of farmers and the unofficial sector in the process of developing local crops. Maintaining biodiversity should be aided by the area's unique agroecology and farmers' preferences for various cultivars with various traits. PPB obviously has the potential to improve current international approaches to plant breeding, but it would be premature to claim that this potential has already been realised at this time. There are still many significant questions like will varieties created by PPB be different from those created through traditional breeding techniques? Will they be better if they differ? Will the extra benefits outweigh the extra expenditures PBB implies if they are better and different? How precisely will PBB's benefits and expenditures be allocated? Will adopting PBB techniques be beneficial to institutional plant breeders? Will farmers consider taking part in PPB initiatives to be worthwhile are questions of serious concerns. Concrete data will be required to show that participatory breeding techniques can live up to the expectations of PPB proponents before the effectiveness of PPB is definitively proved, but additional research is needed to clear many remaining uncertainties. Even if PPB results in the development of varieties that are not only superior to conventionally bred varieties but also different from them, the development of these varieties is probably going to result in additional expenditures for both plant breeders and end users, thus including them only at a stage where they could impart a significant difference. Ultimately, participatory plant breeding is able to combine the maintenance and the enhancement of agro-biodiversity with the need to feed everybody by making the food available and accessible, and with the need to cope with a continuous and gradual change in the climatic conditions

REFERENCES

Bhargav DK, Meena H and Patil PPB. 2014. Participatory plant breeding: farmers as breeders. *Popular Kheti* 2: 7-14

Ceccarelli S. 2012. Plant breeding with farmers. A technical manual. ICARDA, Aleppo, Syria

Ceccarelli S. 2016. Participatory barley breeding in Syria. Farmers' Crop Varieties and Farmers' Rights: Challenges in Taxonomy and Law, 84

Enjalbert J, Dawson JC, Paillard S, Rhoné B, Rousselle Y, Thomas M and Goldringer I. 2011. Dynamic management of crop diversity: From an experimental approach to onfarm conservation. *Comptes rendus biologies* 334: 458-468

Halewood M, Deupmann P, Sthapit B, Vernooy R and Ceccarelli S. 2007. Participatory plant breeding to promote Farmers' Rights. *Bioversity International*, Rome, Italy. 7 pp.

Merga W. 2017. Review on Participatory Plant Breeding. *International Journal of Research Studies in Agricultural Sciences* 3: 9 http: //dx.doi.org/10.20431/2454-6224.0309002

Morris ML and Bellon MR. 2004. Participatory plant breeding research: Opportunities and challenges for the international crop improvement system. *Euphytica* 136: 21–35

Rahman M, Thant A, Win M, Tun M, Moet P, Thu A, Win K, Myint T, Myint O and Tuntun Y. 2015. Participatory varietal selection (PVS): a" bottom-up" breeding approach helps rice farmers in the Ayeyarwady Delta, Myanmar. *SABRAO Journal of Breeding and Genetics* 47

Shelton AC and Tracy WF. 2016. Participatory plant breeding and organic agriculture: A synergistic model for organic variety development in the United States. *Elementa: Science of the Anthropocene* 4: 000143 doi: 10.12952/journal.elementa.000143

Singh S, Gaurav SS and Singh SP. 2019. PARTICIPATORY PLANT BREEDING: AN OVERVIEW. International Conference on Global Research Initiatives for Sustainable Agriculture and Allied Sciences (GRISAAS-2019), ICAR-National Acadamy of Agricultural Research Management, Hyderabad, Telangana (India)

Singh Y, Nayak A, Sharma D, Gautam R, Singh R, Singh R, Mishra V, Paris T and Ismail A. 2014. Farmers' participatory varietal selection: A sustainable crop improvement approach for the 21st century. *Agroecology and sustainable food systems* 38: 427-444

Sthapit B, Joshi K, Gyawali S, Subedi A, Shrestha K, Chaudhary P, Rana R, Rijal D, Upadhaya M and Jarvis D. 2002. Participatory Plant Breeding: Setting Breeding Goals and Choosing Parents for On-Farm Conservation Strengthening the Scientific Basis of In Situ Conservation of Agricultural Biodiversity On-Farm: Nepal Country Component, which is funded by the Directorate-General for International Cooperation, the Netherlands.

Chapter 10

Special Breeding Methods: Reverse Breeding

M. Vennela and D. Dinesh Varma*

Department of Genetics and Plant Breeding, Dr. Rajendra Prasad Central Agricultural University, Pusa, Bihar
****e-mail: mudavathvennela235@gmail.com***

Introduction

The discovery that hybrid progenies often outperform their homozygous parents in terms of size, growth traits, and yield-a phenomenon known as heterosis-one of the most crucial discoveries in plant breeding. Since it is difficult to predict which paternal lines would produce the best offspring, hybrid breeding has historically mostly relied on trial and error. The offspring of numerous parent pairings must next be crossed and evaluated. The difficulty of reproducing uncharacterized heterozygotes by seeds in present plant breeding programmes prevents significant levels of variety from being attained. Due to the segregation of characteristics, the elite heterozygote's advantageous allele combinations are lost in the following generation. One method for creating homozygous parental lines from heterozygous plants is reverse breeding. It's the revolutionary technology pioneered by Rob *et al.* (2009).

Heterozygous germplasm, which have best performance ability is chosen for further breeding after these populations have been evaluated under various environmental circumstances. By selecting heterozygous plants with desirable traits, this procedure creates homozygous lines that, when crossed, can replicate the original genetic make-up of the chosen heterozygous plants. In this sense, the reverse breeding process aids in developing hybrids in greater quantity with less effort than traditional plant breeding techniques. It is no longer necessary to determine the origin of unidentified heterozygous plants in order to repair the genomes in immortal lines.

According to the theory, reverse breeding permits the reshuffling of chromosomes between two homozygous plants. Some genetic alterations leave the shuttle residues, such as bacteria and fungus, in the host plant, while novel breeding techniques, such as reverse breeding, enable the development of homozygous lines without the introduction of DNA sequences. Despite the introduction of recombinant DNA during the reverse breeding process, the chosen homozygous parental lines and their progeny are not transgenic.

Reverse breeding studies on Arabidopsis thaliana suggests that this approach could be employed in crop development, however it might be difficult to put into practice in crops with more chromosomes. Commercial CMS lines, which are frequently employed in current agriculture, are perfectly compatible with RB. This strategy only works with crops that can renew DHs from spores and have a haploid chromosomal number of 12 or below. An alternative reconstruction technique has been proposed for polyploid animals or animals with a high number of chromosomes. In the process of reverse breeding, a genetic constitution step is used to prevent genetic recombination, producing intermediate plants that are subject to GMO regulations. However, the final chosen variety and their ancestors don't have this genetic modification, hence they are exempt from the GMO law.

RB could be used to accomplish the following objectives:

- ✰ To create breeding groups for unidentified hybrids.
- ✰ By genetically advancing the parental lines, to increase hybrid performance.
- ✰ To retain a homozygous parental line that has produced a highly heterozygous plant.
- ✰ For the study of gene interactions, these homozygous chromosomal replacement lines offer incomparable resources.

The method not only enables the fixation of uncharacterized germplasm but also gives breeders access to a tool for breeding that, when used on plants with known backgrounds, enables the rapid generation of chromosome substitutions that will facilitate breeding at the level of a single chromosome (Kumari *et al.*, 2018).

Mechanism of Reverse Breeding

1. Heterozygote Selection

Irrespective of whether its parents are known or not, a highly heterozygous plant with a desirable characteristic combination is picked. The heterozygote develops into a gamete.

2. Meiotic Recombination Suppression at the Time of Spore Production

Effective meiotic crossover suppression is necessary for reverse breeding. Consequently, genes that are necessary for crossover creation yet maintain the chromosome structure are especially helpful. Targeting genes with RNA interference (RNAi) will mostly result in post-transcriptional gene silence, which is necessary for RB, to reduce gene expression. As an alternative, the target gene's dominant-negative mutations can be employed (Ankita *et al.*, 2020). The following genes are necessary for meiotic recombination to occur:

1. Disrupted Meiotic cDNA: DMC1 gene
2. Sporulation-specific gene SPO1
3. Recombinase gene RecA A gene

In hybrids of A. thaliana, the DMC1 gene, which encodes the meiotic recombination protein Disrupted Meiotic cDNA1, prevents non-recombined paternal chromosomes from recombining.

Utilizing a chemical agent that stops recombination to speed up the reverse breeding process is another way to decrease meiosis. Chemicals like MIRIN and the Mre11- Rad50-Nbs1 complex inhibitor stop the G2 stage and prevent the phosphorylation of ATM *i.e.*, Ataxia Telangiectasia Mutated (Dupree *et al.*, 2008). The absence of transgenes in the RB products (DHs) produced by the use of chemicals that prevent crossovers is a significant benefit.

Other methods might be used with crops when steady transformation is challenging or impossible. One among them could be Virus-Induced Gene Silencing (VIGS) which is proven to be a useful method for PTGS induction. For RB to benefit, crossover suppression should not be fully suppressed. One residual crossing may still occur in any chromosomal pair, which can be explained. A single crossing results in the regular segregation of the implicated homologues, which doubles the likelihood of producing a balanced gamete. Two recombinant chromatids are also produced by a crossing, but they are useless for RB. The other two chromatids in the bivalent pair, however, are non-recombinant and useful because a crossover only affects half of them.

3. Creation of Double Haploids

It is possible to create doubled haploid plants through achiasmatic meiosis or from microspore and anther cultures. Immature pollen grains develop into cell colonies using tissue culture methods referred to as "anther culture" and "isolated microspore culture". To encourage the growth of shoots and subsequently roots, the colonies are moved to media containing various plant growth regulators and carbohydrates. The effectiveness of DH production from haploid spores depends on the species. The development of RB is restricted to crops where the use of DH technology is widespread. For the vast majority of crop species, this technology is well-established, and reputable breeding businesses regularly employ such methods in their breeding programmes. The DH approach was used to choose viable selfing lines that can create a hybrid with the same genotype as the original parents (Wijnker *et al.*, 2012).

Modern plant breeding routinely uses molecular markers to genotype DHs, which is essential for RB (De Vienne, 2003). One polymorphic molecular marker per chromosome would be sufficient to genotype every DH in the full absence of meiotic recombination since the entire chromosome would function as a single linkage block. Two markers are required per chromosome in the event of any remaining crossovers.

4. Selection of Complementary Lines through Marker Assisted Selection

The complimentary parents are identified and crossed to recreate the original hybrid using Marker Assisted Selection (MAS). One polymorphic molecular marker per chromosome would be sufficient to genotype every DH in the condition of complete deprivation of meiotic recombination because the entire chromosome would behave as a single linkage block. If there are any residual crossovers, however, two markers are needed per chromosome. Because the hybrid created through RB does not contain a transgene, they should not be categorized as genetically modified (GM).

Revers Genetics Approaches

Investigating the effects of generated variation within a particular gene and determining gene function are the two main objectives of reverse genetics. Gene silencing and homologous recombination are two examples of targeted disruption or change processes. Non-targeted random disruptions include chemical mutagenesis and transposon mediated mutagenesis. Targeted gene silencing by RNA interference and TILLING (Targeting Induced Local Lesions in Genomes) are two important methods of reverse genetics.

- ☆ **RNA interference:** Small double-stranded RNA (dsRNA) molecules are used as triggers in the majority of eukaryotic cells' RNA interference (RNAi) regulation mechanism, which involves homology-dependent control of gene activity. An exciting approach for reverse genetics is RNAi-based silencing (Waterhouse *et al.*, 1998). Recent advances in RNA interference have made it possible to analyse genes' loss-of-function phenotypes and mutant their expression, enabling analysis of gene function in the absence of mutant alleles. RNAi has demonstrated to be applicable to many organisms and has been used to produce a wide range of loss-of-function phenotypes since it has been shown to function similarly in all metazoans (Aklilu, 2021).

The mechanism is started by the enzyme Dicer, which breaks lengthy dsRNA molecules into short double-stranded pieces called small interfering RNAs (siRNAs), which have between 21 and 23 nucleotides. The sense strand and the guide (antisense) strand of each siRNA are unwound to form two single-stranded RNAs (ssRNAs), respectively. The protein Argonaute 2 then cleaves the passenger strand (Ago2). The guide strand is integrated into the RNA-induced silencing complex (RISC) while the sense strand is destroyed. The target mRNA is then bound and degraded by the RISC assembly. This is done specifically when the guide strand interacts with a complementary sequence in an mRNA molecule and triggers the RISC enzyme Ago2 to cleave the mRNA.

☆ **TILLING:** Targeting Genome-Induced Local Lesions (TILLING) is a generic reverse genetic method that identifies point mutations in regions of interest by combining chemical mutagenesis with PCR-based screening (McCallum *et al.*, 2000). Using a mismatch-specific endonuclease, this method can identify induced or uninduced DNA polymorphisms in target genes. Finding an allelic series of induced point mutations in genes of interest is made easier with the aid of a recently discovered comprehensive reverse genetic method. It makes it possible to quickly and affordably identify induced point mutations in populations of people who have been physically or chemically altered.

The method entails amplifying the target gene via PCR using fluorescently labelled primers, forming a DNA heteroduplex between wild type and mutant alleles (PCR products corresponding to the mutant and wild type alleles are heated and then slowly cooled), and then performing endonuclease digestion that specifically cleaves at the site of an EMS-induced mismatch. By alkylating G residues, which pair with T instead of the conservative base pairing with C, EMS creates transitional mutations (G/C, A/T).

Marker Assisted Reverse Breeding

Reverse breeding is a flexible approach, especially when used in conjunction with high throughput genotyping. Evidently, early-stage populations of DHs can identify complementing parents more quickly with the help of high throughput genotyping. Researchers can study gene interactions in the heterozygous inbred families that can be produced by crossing and backcrossing reverse breeding products. Fast QTL discovery is possible when genotyping is combined with other analyses, such as transcriptome or metabolome analysis, to screen populations for traits on a single chromosome that segregate. It is also simpler to generate chromosome-specific linkage maps and fine-map genes using these heterozygous inbred families. Reverse breeding can therefore be incredibly helpful in learning more about the type of heterotic effect. Any maize hybrid can be turned into inbred lines with any necessary degree of resemblance to its original parent lines using marker-assisted reverse breeding (MARB). In comparison to conventional pedigree breeding utilising hybrids, this method offers the benefits of quick speed (one to two years), fixed heterotic mode, and swift recovery of over 80 per cent of the parental genotypes and related advantageous genes (Guan *et al.*, 2015).

General protocol for marker assisted reverse breeding:

1. Take the DNA from the pericarp and seed embryo of a particular elite hybrid separately.
2. Choose a genotyping platform with molecular markers that offer high levels of genome coverage and high throughput genotyping.
3. To determine the parental genotypes, genotype the DNA samples from the pericarp and seed embryo.
4. For ensuing marker-assisted selection, choose polymorphic markers that determine the parental genotypes.
5. Self the hybrid F1 to produce F2 seeds, then genotype the F2 seeds or plants using the subset markers to determine which offspring are most similar to their respective maternal and paternal genotypes.
6. Continue selecting among the F3 seeds or plants after selfing the chosen F2 plants to produce F2 derived F3 families.
7. After selecting among F4 seeds or plants to find the offspring that are most similar to their maternal and paternal genotypes, respectively, self the selected F3 plants to produce F3-derived F4 families.
8. Continue marker-assisted selection up until the chosen progeny achieve a desired level of resemblance to the parental lines before moving on to the following stage.
9. Get fixed genotypes by using DH technology or keep selfing.

Applications

1. **Reconstruction of heterozygous germplasm**: RB can hasten the establishment of varieties for crops that don't yet have a large pool of breeding lines. Superior heterozygous plants in these crops can be propagated without the genetic makeup of the parent plant being known in advance. A heterozygous F1 is produced by crossing two homozygous parents (grey and black bars). Selfing F1 results in the production of a segregating F2 population. A start hybrid with an

unknown genetic makeup is chosen for its desirable traits and put through the two reverse breeding phases (grey box). Meiotic crossing over is prevented, allowing complete parental chromosomes to be transmitted by spores without rearrangement. The DH lines are then created utilising in vitro culture procedures using the achiasmatic gametes. Complementary parents can be chosen from these populations, that when crossed will reconstitute the original hybrid (Mandeep *et al.*, 2019).

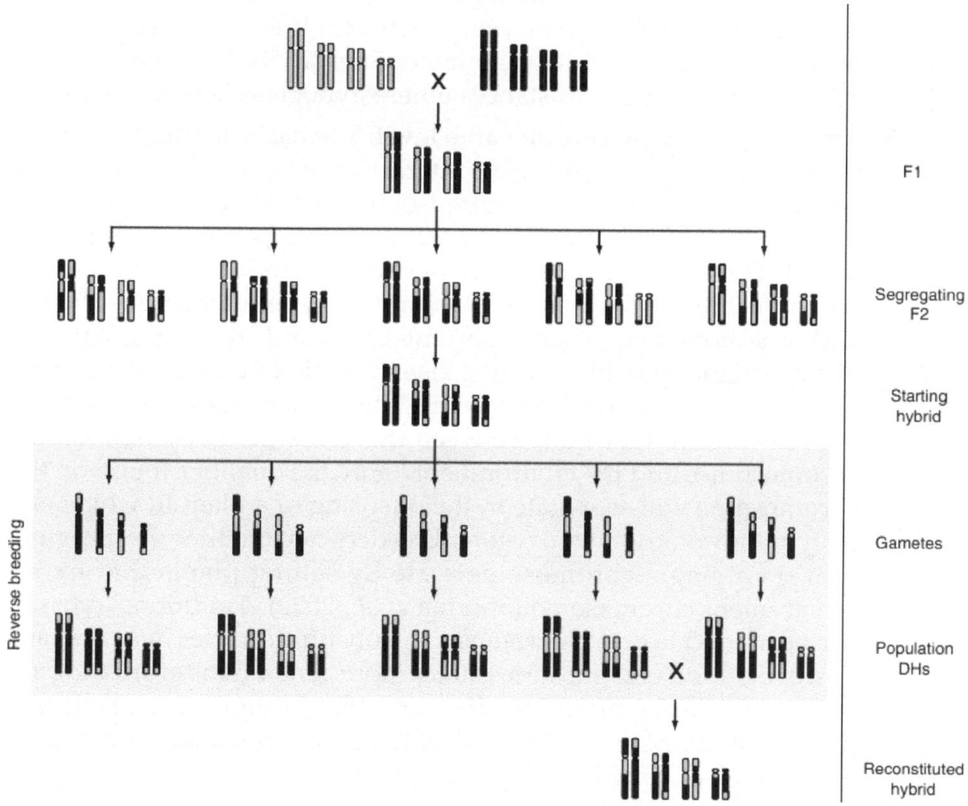

Figure 10.1: A Heterozygous F1 is Produced by Crossing Two Homozygous Parents (Grey and black bars).

The F1 population creates a segregating F2 population when selfed. A start hybrid with an unknown genetic makeup is chosen for its desirable traits and put through the two reverse breeding phases (grey box). Meiotic crossing over is prevented, allowing complete parental chromosomes to be transmitted by spores without rearrangement. The doubled haploid (DH) lines are then created from the achiasmatic gametes utilising in vitro culture procedures.

2. **Transfer of CMS**: According to this theory, a CMS donor is rendered homozygous in order to confer recombination suppression in order to

create a line that is identical to a desirable inbred or pure line but with a CMS background. By pollinating the aforementioned homozygous recombination inhibited line with the pollen of the target line, a first cross is created. Breeders exploit cytoplasmic male sterility in a number of vegetable crops, including cabbage and carrots (CMS). The existence of male sterility in these systems poses a unique challenge to RB. In such circumstances, DH plants can be produced via gynogenesis rather than androgenesis. In that the chromosomes from the maintenance line may be retrieved directly in the cytoplasm of the sterile line in a single step, this is completely consistent with RB. It is conceivable to cross male sterile (A) lines with maintainer lines (B) that carry one copy of a restorer gene in circumstances where gynogenesis is inadequate.

3. **Breeding on single chromosome level**: The basis of many intriguing traits in plants is polygenic gene interaction, which is frequently found on many chromosomes. Chromosome substitution line is created when it is applied to an F1 hybrid of known parents, as explained by RB. These homozygous lines of chromosome substitution enable a new technique for the investigation of gene interaction. Hybrids with one homozygous chromosome can result by mating with one of the original parents, but it is also possible to create hybrids with just one heterozygous chromosome. The former enables research into the structural interactions between the genes given by replacement chromosomes and the environment. Only the qualities found on that chromosome will segregate in the offspring of a plant in which only one chromosome is heterozygous. Breeders can fine-tune the intriguing trait on a single chromosome scale by selfing plants that have a replacement chromosome (Sharma *et al.*, 2022). Traditional crossing has been used to create chromosome-substitution lines in A. thaliana, however, this technique necessitates more generations of crossing and intensive genotyping to identify a specific chromosome-substitution line. With reverse breeding, all potential chromosome-substitution lines can be obtained quickly.

Advantages

- Reverse breeding's key benefit is that it makes it easier to choose superior hybrid plants. Without knowing their genetic makeup beforehand, huge numbers of plants can be produced, tested, and well-performing plants can be continuously reproduced.
- By accelerating the breeding process and expanding the genetic combinations that are accessible, RB enables breeders to produce superior varieties much more quickly in response to farmer demands.
- This essentially eliminates the random components present in earlier hybrid breeding.

- The method will be crucial for crops that lack extensive collections of existing breeding lines.

Limitations

There are many factors that must be examined for effective performance of any technique and considering this even reverse breeding have its own limitations.

- Only crops where the use of double haploid technology is widespread can use this strategy.
- The crops in which double haploids are barely formed include soybean, lettuce, cotton, cucumber, broccoli, maize, sorghum and tomato.
- It can only be used in crops with haploid chromosomal numbers of 12 or fewer, or in crops where spores can regenerate into DHs.
- The presence of non double haploid plants required to find the complementary pair that recreates the original heterozygosity in plants with more chromosomes would be exceedingly high and practically impossible (Lusser *et al.*, 2012).
- There is no room for additional selections due to the total homozygosity of the obtained plants, which restricts the desired genetic variation for plant breeding.

Conclusion

Any desired heterozygote can be chosen from a huge population and perpetuated eternally as F1 by reverse breeding. This is significant because it addresses one of the drawbacks of conventional breeding, which produces hybrids by carefully regulating the crossing of a small number of founder lines. Without knowing the origin of the plants, the genomes of uncharacterized heterozygous plants can now be fixed. It aids in limiting the breakdown of complicated genotypes into homozygous parental lines and permits further development of these lines using conventional breeding techniques. The regeneration of haploid spores into DHs combined with crossover suppression yields new and effective breeding applications. Commercial CMS lines that are routinely employed in contemporary agriculture are entirely compatible with RB. While it is primarily only utilised in crops with fewer chromosome number, this approach offers the potential to revolutionise genetic engineering and selective breeding in agriculture.

Future Potential

- RNAi Mediated Reverse Breeding is a new technique that needs further research to solve technical issues.

- ☆ Future agricultural yield may be improved by new opportunities for RB to choose and enhance advantageous genotypes.
- ☆ To increase DH production efficiency, more research is necessary.
- ☆ Since RB may make it possible to create parental lines for the re-creation of hybrids, its potential for improving agricultural crops could be foreseen.

REFERENCES

Aklilu, E. 2021. Review on forward and reverse genetics in plant breeding. *All Life*. 14 (1): 127-135.

Ankita, S., Ravi, S., Ujjwal, K., Kumari Rekha., Apoorva, P and Ashwini, K. 2020. Reverse breeding: A modern plant breeding approach for hybrid recreation. *International Journal of Chemical Studies*. 8 (3): 1128-1131.

De Vienne, D. 2003. Molecular Markers in Plant Genetics and Biotechnology. Enfield, NH, USA: Science Publishers, Inc

Dupree, A., Boyer-Chatenet, L., Sattler, RM., Modi, AP., Lee, JH and Nicolette, ML. 2008. A forward chemical genetic screen reveals an inhibitor of the Mre11–Rad50–Nbs1 complex. *Nature chemical biology*. 4 (2): 119.

Guan., Yi-Xin., Wang, BH., Yan, F and Ping, L. 2015. Development and application of marker-assisted reverse breeding using hybrid maize germplasm. *Journal of Integrative Agriculture*. 14 (12): 2538–2546.

Kumari, P., Nilanjaya and Singh, N. K. 2018. Reverse breeding: Accelerating innovation in plant breeding. *Journal of Pharmacognosy and Phytochemistry*. 1: 1811-1813.

Lusser, M., Parisi, C., Plan, D and Rodriguez-Cerezo, E. 2012. Deployment of new biotechnologies in plant breeding. Nature biotechnology. 30 (3): 231.

Mandeep, K., Harshal, A., Avinashe, Nidhi, D., Kesha., Ram., Simerjeet, K and Swapnil, K. 2019. Reverse Breeding: Creating Parental Line for a Heterozygous Plant and its Complication. Annals of Biology. 35 (1): 50-54.

McCallum, C. M., Comai, L., Greene, E.A and Henikoff, S. 2000. Targeted screening for induced mutations. *National Biotechnology*. 18: 455-457.

Rob, D., Kees, V. D., Bastiaan de Snoo, Mark, V. den Berg, Cilia L. C. Lelivelt, William V., Leo Woudenberg, Jack, P. C. de Wit, Kees, R., Johan, W. S., Eveline van der Zeeuw, Aat Vogelaar, Gerald, F., Evert, W., Gutteling, Marina, N. Keppel, Paul, van Drongelen, Matthieu, K., Philippe, E., Alisher, T., Hong, Ma, Hans de Jong and Erik, W. 2009. Reverse breeding: a novel breeding approach based on engineered meiosis. *Plant Biotechnology Journal*. 7 (9): 837- 845.

Sharma, M., Vasisth, P., Vaibhav, C and Prasad, H. 2022. Reverse breeding: A novel plant breeding technique. *Biotica Research Today*. 4 (6): 435-437.

Waterhouse, PM., Graham, MW and Wang, M. 1998. Virus resistance and gene silencing in plants can be induced by simultaneous expression of sense and antisense RNA. *Proceedings of the National Academy of Sciences*. 95: 13959–13964.

Wijnker, E., Van Dun, K., De Snoo, CB., Lelivelt, CL., Keurentjes, JJ and Naharudin, NS. 2012. Reverse breeding in Arabidopsis thaliana generates homozygous parental lines from a heterozygous plant. *Nature Genetics*. 44 (4): 467.

Chapter 11

Breeding for Crop Quality

Madhuri Pradhan

Assistant Professor, C.V. Raman Global University, Bidyanagar, Mahura, Janla Bhubaneswar – 752 054, Odisha
e-mail: madhuri.pradhan@cgu-odisha.ac.in

Introduction

The quality of agricultural yield has evolved into a vital aspect. The appropriateness of an economic plant product in connection to a particular application or goal is what is meant to be understood by the term "quality." The preference of customers is significantly impacted by the product's quality. The concept of quality encompasses a number of characteristics of an economic good or service.

Quality Traits

Quality traits differ from crop to crop depending upon the plant part used as an economic product.

The following is a list of the desirable quality qualities that can be found in various crops:

i. Rice

In India, white coloured and long slender grains are mostly preferred. In general, aromatic, less breakage of grain during milling, more hulling recovery, good cooking quality (depends chiefly on amylose content and gelatinization temperature), high protein and lysine contents are notable quality traits.

ii. Maize

In maize, bold flint grains with yellow or white colour, increased lysine and tryptophan content, oil are important quality traits. Popping expansion in popcorn and high sugar content in sweet corn are also examples of quality traits.

iii. Wheat

Grain sizes that are bold and robust, grain textures that are firm, vitreous, and lustrous, and bold grain sizes. A significant contribution to the quality of the protein is made by a high lysine level. Important qualities for creating dough include a high gluten content and a sedimentation value of between 20 and 40. The pelshinke value of wheat is what determines whether or not it will be used to make chapati, bread, biscuits, or confectionery.

iv. Pearl Millet

In pearl millet, the most desirable characteristics include bold, lustrous, amber-colored grains; palatability; high iron content; high crude fibre content; and low erucic acid content.

v. Barley

In barley, plump grain, high protein and lysine content and high extract of soluble oligosaccharides after malting are preferred quality traits.

vi. Sorghum

It is preferable for sorghum to have a higher amount of lysine and protein in its white grains and vitreous endosperm.

vii. Pulses

When it comes to pulse crops, desirable characteristics include seed size and colour of grains, as well as high protein content (21–27 per cent of total), as well as methionine, tryptophan concentration, and digestibility.

viii. Oil Seeds

In oil seed crops, desirable characteristics include seed size and colour, high oil content, absence of anti-nutritional elements (low erucic acid, low glucosinolate content, and low fibre), and a higher proportion of unsaturated fatty acids (higher iodine number).

ix. Tobacoo

In tobacco, the preferred leaves for cigars are those that are short and thin, with less branching veins. While smoking from a pipe, you should look for leaves that are rather thin. For making cigarettes, thicker leaves are preferable. It is preferable to have a high nicotine concentration in bidis, hookahs, and chewing tobacco, while having a low nicotine amount in cigarettes. A high concentration of sugar is also recommended.

x. Potato

Characteristics that are desirable in potatoes include their shape, size, colour, and texture of tubers, as well as their taste, quality of cooking, thin skin, quality of keeping, and high starch content.

xi. Cotton

The most desirable characteristics in cotton fibre include its length, strength, fineness (ranging from fine to extremely fine), maturity (ranging from good to very high maturity), uniformity, and white colour.

xii. Vegetables

When it comes to vegetable crops, some of the most noticeable desirable characteristics include the fruit's size, shape, colour, high vitamin and mineral content, pleasant flavour, maintaining quality, and processing quality.

Major breeding objectives for quality cops in accordance with the requirements of the different crop plants:

1. High protein content and quality,
2. High oil/fat content and quality,
3. High vitamin contents and mineral nutrients, and
4. Low content of toxic and harmful substances.

Nutritional Quality of Crops

Nutrition and Nutrients

Nutrition refers to the multidisciplinary field of research that examines the effects of diet on human health. Food is the substance that is consumed in order to meet the requirements of the body for the generation of energy, the preservation of health, and the promotion of growth and reproduction. Nutrients are a broad term that refers to a number of different chemical components of food that are responsible for supplying the body with nourishment. Carbohydrates, lipids, proteins, vitamins, minerals, and water are the components that make up a balanced diet. These components are required by the body on a consistent basis and in a predetermined amount for optimal health. The term "good nutrition" refers to consuming a enough amount of a healthy, well-balanced diet that provides the body with all of the vital nutrients it needs. A low food intake is what is meant by the term "poor nutrition." A lack of nutrients, an excess of nutrients, or an imbalance of nutrients can all lead to malnutrition. Because of this, it is important to consume sufficient amounts of each nutrient.

Proteins

Amino acids are the fundamental building blocks of proteins, which are chemical molecules. A carboxyl (COOH), also known as an acidic group, and an amino (NH2), sometimes known as a basic group, are found in each amino acid. The diets of humans require a certain amount of each of the 21 different amino acids. They are able to be separated into two distinct categories, namely:

1. Essential amino acids and
2. Non-essential amino acids.

The human body is unable to produce the essential amino acids, hence the only way to fulfil the body's demand for them is through the consumption of food. There are 10 different amino acids that are regarded as being important for infants. Methionine, isoleucine, leucine, lysine, threonine, tryptophan, valine, phenylalanine, histidine, and arginine are the amino acids that make up this group. Arginine and histidine, on the other hand, are amino acids that are not considered essential for adults. The human body is able to produce the non-essential amino acids on their own, therefore they do not have to be obtained through food consumption. Cystine, cysteine, proline, glycine, serine, alanine, aspartic acid, hydroxyproline, glutamic acid, norleucine, and tyrosine are some of the amino acids that fall into this category. The amount of necessary amino acids that are present in a protein will indicate its overall quality. The amino acids that include sulphur and are known as limiting amino acids are tryptophan, threonine, isoleucine, lysine, valine, and methionine (Table 11.1).

Table 11.1: List of Essential Amino Acids Deficient in different Crop Plants

Food	*Deficient Essential Amino Acids*
Cereals	Lysine, Tryptophan, Threonine
Pulses	Methionine, Tryptophan
Nuts and oilseeds	Lysine
Green-leafy vegetables	Methionine

Vitamins

Vitamin refers to a critical substance that is required in very small quantities but is essential for normal growth and good health. Vitamins are essential for appropriate growth and good health. Vitamins have a crucial role in the expansion, maintenance, and reproduction processes of the body. There are two types of vitamins, one that dissolves in fat (A, D, E, and K) and another that dissolves in water (B, C, and some of the B vitamins) (B group vitamins and vitamin C). Both vitamin D and folic acid are capable of being partially produced by the body itself. The remainder of the vitamins must come from the food that we consume. Fats are a source of vitamin D, and green vegetables and fruits with meaty centres are good sources of other vitamins.

Nutritional Quality of Cereals and Pulses

A significant proportion of carbohydrates can be found in cereals. Cereals are responsible for 70 per cent of the total number of calories that are consumed by humans. They contribute fifty percent to the world's requirement for protein, whereas grain legumes only contribute twenty percent. The remaining thirty percent originates from various animal products. The cereals have a protein level that ranges on average from 10 to 13 per cent of their total weight. Yet, there is a lack of the amino acids lysine, tryptophan, and threonine in the cereal protein. Four types of proteins are found in cereals as given below:

1. **Albumins:** Soluble in water.
2. **Globulins:** Soluble in saline solutions.
3. **Prolamins:** Soluble in relatively strong alcohol.
4. **Glutelins:** Soluble in dilute alkaline solutions.

The cereals we eat all have prolamins that are quite high in levels of the amino acids proline and glutamine, but they have low levels of the basic amino acids, including lysine. The presence of prolamin in cereals has been shown to have an inverse relationship with the amount of total protein they contain.

About seventy percent of cereals are composed of glutelin and prolamin, whereas the remaining thirty percent is albumin and globulin. The most major forms of protein found in vegetables are grains and legumes. They contain between 18 to 28 per cent protein, which is almost double the amount found in grains. Albumin and globulin are two distinct kinds of proteins that can be discovered in pulses. Globulin, which accounts for around 80 percent of the seed's total protein content, is the grain legumes' primary source of storage protein.

There are three different forms of globulin found in chickpeas: alpha, beta, and gamma. A greater percentage than eighty percent of the total protein is comprised of alpha globulin. Legumin and vicilin are the two varieties of globulin that may be found in faba and pea, respectively. Vicilin dissolves more easily in a salt solution than legumin does. In addition, in contrast to vicilin, legumin does not coagulate when exposed to high temperatures (95 degrees Celsius).The amino acids methionine and tryptophan are in short supply in bean proteins.

In addition, they include a number of compounds that are known to be hazardous, such as protease inhibitors, haemagglutinins, lathyrogens, glucocides, goitrogens, cyanogens, metal binding factors, and antivitamin factors. Cooking, on the other hand, eliminates a significant portion of these carcinogenic chemicals. It will be necessary to eradicate some of these poisons through selective breeding. The consumption of grains and pulses in a proportion of three to one results in a diet that is nutritionally sound.

Genetics of Nutritional Traits

Inheritance of nutritional quality characters may be governed by:

1. Oligogenes with distinct individual effects,
2. Polygenes with small additive effects, and
3. Cytoplasmic genes

1. Oligogenic Inheritance

Oligogenic inheritance is the term used to describe situations in which the inheritance is controlled by only a single or a small number of important genes. With this form of inheritance, each gene has a sizable effect on the manifestation of the nutritional quality character, and this effect is easy to discern. Characters of high value can be easily differentiated from those of low value due to their obvious differences. In sorghum, a single gene with partial dominance is responsible for controlling the high lysine content. In barley, the presence of a high lysine content is controlled by one major gene in addition to many minor genes. One primary gene in safflower is responsible for determining the fatty acid composition, and this gene has three major alleles. Two primary genes, in addition to other modifiers, are responsible for the high beta-carotene concentration in tomato.

2. Polygenic Inheritance

Polygenic inheritance is a situation in which the inheritance is controlled by a number of genes, each of which has a very minor additive influence. When a trait is passed down in this manner, the range of possible expressions for that trait is unbroken from one extreme to the next. Because of this, it is not possible to classify plants into easily discernible categories. Polygenes are responsible for determining the amount of protein found in cereals and pulses as well as the amount of seed oil found in oilseed crops. Characteristics that are controlled by polygenes are susceptible to the effects of changes in their environment and, in general, have a low heritability. A complicated carotenoid inheritance pattern can be seen in carrots due to their high carotenoid content. Anti-nutritional compounds such as erucic acid and eicosenoic acid are controlled by oligogenes in mustard and rapeseed. These oligogenes have many alleles. Both the maternal parent and the seed genotype play a role in determining the amount of oil that is contained within the seeds of maize and rapeseed. In a similar vein, the genotype of the maternal parent can have an effect on the fatty acid composition of maize, soybeans, flax, and rapeseed. The genotype of the seed can, however, have an effect on the content of the fatty acids found in flax, rapeseed, and safflower.

Sources of Quality Traits

Improved quality can come from one of these places:

1. Cultivated varieties of crop,

2. Germplasm collections,
3. Wild relatives and species
4. Mutants
5. Somaclonal variants
6. Transgene

Cultivated varieties of crop are the most common. In addition to this, mutations are significant contributors to the overall improvement in nutritional quality. When it comes to agricultural plants in general, cultivated varieties and collections of germplasm are major sources of increased nutritional quality features. A significant amount of the high lysine content found in sorghum, maize, and barley can be attributed to mutations, both spontaneous and induced (Table 11.2).

Table 2: List of the Mutants Affecting Protein Quality in Cereals

Crop	*Mutant*	*Mutant type*	*Protein quality*
Maize	Opaque-2	Spontaneous	High Lysine, Tryptophan
	Opaque-7	Spontaneous	High Lysine, Tryptophan
	Floury-2	Spontaneous	High Lysine and S-amino acids
	Brittle-2	Spontaneous	High Lysine, Tryptophan
Barley	Hiproly	Spontaneous	High Protein, Lysine
	Notch-1	Ethyl methane Sulphonate	High Protein, Lysine
	Notch-2	Ethyl methane Sulphonate	High Protein, Lysine
	Riso 1508	Ethyl imine	High Lysine
Sorghum	IS11167	Spontaneous	High Lysine
	IS11758	Spontaneous	High Lysine
	P721 opaque	Diethyl sulphate	High Lysine
	P721 vitreous	Diethyl sulphate	High Lysine

A high level of unsaturated fatty acids has been transferred from wild species to the farmed oil palm, which is known scientifically as *Elaeis guineensis* (*E. oleifera*). It is preferable to have lower levels of linolenic acid in soybeans. Yet, wild species have a higher concentration of linolenic acid than cultivated species do. As a result, wild species are not suitable for usage in this context. The high levels of beta carotene seen in tomato (*Lycopersicon esculentum*) were introduced from wild green-fruited species (*L. hirsutum*).

Breeding Methods

There is no distinction between the breeding techniques utilised for the enhancement of quality and those utilised for the improvement of any other character. Backcrossing, pedigree procedures, single-seed descent, recurrent selection, progeny selection, and mutation breeding are only some of the other breeding techniques that are utilised widely for the purpose of improving quality features. The protein content of the common bean, also known as Phaseolus vulgaris, has been improved through the application of the pedigree, single seed descent, and recurrent selection breeding procedures. The amount of seed protein rose from 21.9 per cent to 24.6 per cent after two cycles of recurrent selection. The amount of seed protein in soybeans increased from 42.8 percent to 46.1 percent after five cycles of recurrent selection. In Sorghum, the development of high-lysine lines was accomplished through the application of the pedigree breeding process. The backcrossing strategy was utilised in barley to facilitate the establishment of high-lysine lines. Barley lines with high levels of lysine produced smaller seeds and lower grain yields. Using the use of modified recurrent selection, Russian breeders were able to enhance the seed oil content of sunflowers from 30 to approximately 50 percent in just 50 years. With a reduction in the amount of hull, the oil content of safflower was able to be boosted from 37 per cent to 50 per cent. During 70 cycles of recurrent selection in the United States, the seed oil content of maize grew from 4.7 to 17 per cent, while the protein content climbed from 10.9 to 23.5. Nonetheless, there was a significant drop in the amount of grain produced. The establishment of high-lysine lines in Sorghum, barley, and maize has been accomplished through the application of mutation breeding. For the purpose of inducing high lysine mutants in sorghum, diethyl sulphonate (DES) has been utilised, whereas in barley EMS and EI have been used.

Screening Techniques

The use of a great deal of chemical analysis is involved in breeding for improved nutritional quality. The breeding material needs to be analysed for its levels of protein and amino acids, seed oil and fatty acid composition, vitamin and anti-nutritional factor content, and anti-nutritional factor levels. In order to accomplish this, biochemists need to work together closely. Chemical analysis is used to choose plants that have a higher overall nutritional quality than other options. The reliability of the tests will determine how successful the breeding effort will be. The procedures for carrying out chemical analysis ought to be straightforward, low-cost, and speedy. There are now technologies available for the quick chemical examination of proteins, seed oils, fatty acids, and amino acids respectively.

Seed Oil Analysis

Nuclear Magnetic Resonance (NMR) or Nuclear Infrared Analyzer are two instruments that are utilised in the process of analysing seed oil (NIR). This method of analysing oil does not involve any damaging processes. Following the completion of an oil analysis, the seeds are ready for planting. It is possible to analyse several single plants from different populations that are segregating. This approach is not only straightforward but also very quick and precise. It is simple to analyse between 300 and 400 samples using this procedure every single day.

Protein Analysis

The use of a protein analyser has made it possible to perform analysis on the protein content of grain. The measurement of infrared reflectance serves as the foundation for this investigation's methodology. The seeds are then ground into flour, which is then used for the examination. This procedure is quite quick. Using this approach, it is simple to analyse between 200 and 300 samples every day. The concentration of each individual amino acid can be determined using either a colorimetric or a microbiological approach. Chemical methods are utilised in the investigation of elements that are detrimental to nutritional value. Although they were more precise, the older methods of determining the amount of protein (the Kjeldahl method) and the amount of oil (the Soxhlet method) required a significant amount of time. Recently, however, procedures that are quick have been discovered.

Breeding for Low Toxic Substances

Toxic chemicals can be found in certain grains, legumes, oilseeds, vegetables, fruits, and forage crops. These hazardous compounds have the potential to have detrimental impacts on the health of both humans and animals. Animals will suffer negative effects to their health if they are given harmful substances in their feed. Thus, it is of the utmost importance to design kinds of forage and food crops that contain a low amount of poisonous chemical so that it will not have a negative effect when consumed by animals.

There has been significant headway made in the production of types of above-ground crops that have low levels of their respective toxins. A great deal of chemistry research is necessary for breeding with the goal of reducing the amount of hazardous compounds. As a result, the development of methods of chemical analysis that are straightforward, low-cost, quick, and trustworthy is vital.

Practical Achievements

When it comes to food, fodder, and vegetable crops, one of the most significant goals of plant breeding is to improve the crop's nutritional content through breeding. In addition to a low level of anti-nutritional components, the important nutritional characteristics include high levels of protein, oil, and vitamin content. Through a number of different nations' efforts, numerous types of food crops have been bred to produce varieties with enhanced nutritional value. The seed protein content of the common bean has increased from 21.9 to 24.6 per cent, whereas the seed protein content of the soybean has increased from 42.8 to 46.1 per cent. The seed oil content of sunflowers in the USSR has gone from 32 per cent to roughly 50 per cent, while the seed oil content of safflowers has climbed from 37 per cent to 50 per cent. The percentage of seed oil in maize has increased from 4.7 to 17 per cent, and the percentage of protein has increased from 10.9 to 23.5 per cent. There have been three high lysine mutants found in sorghum (IS-11167, IS-11758, and P-721 opaque), four high lysine mutants found in maize (Opaque 2, Floury 2, Opaque 7, and Brittle 2), and four high lysine mutants found in barley (Hiproly, Notch-1, Notch-2, and Riso-1508) These high-lysine sources are being utilised in an effort to increase the amount of lysine that is present in the targeted crop. Using the Opaque-2 mutant, three composite maize cultivars with high lysine content - Protina, Shakti, and Rattan have been produced for commercial cultivation in India. These maize cultivars include Protina, Shakti, and Rattan. In oilseeds, several cultivars with high seed oil content have been released. In the field of sugarcane, India has seen the emergence of a number of high-sucrose cultivars for use in commercial agriculture (Table 36.5). In wheat, Atlas-66 is a significant source of high protein that is utilised in breeding programmes to boost wheat's protein content. These programmes aim to increase the grain's overall protein content. Because it has a high lysine and protein content, the Indian strain Naphal is being utilised in wheat breeding programmes to boost both the lysine and seed protein content of the crop. Sugarcane varieties with a high soursop content include CO 671, CO 6806, CO 7314, CO 7704, and CO 62174.

Limitations

Breeding for higher nutritional quality is an important part of the solution to the problem of malnutrition, notably for vegetarian people but also for animals. This is especially true for people.

The success of breeding for improved nutritional quality is determined mostly by three different aspects. These factors are as follows: (1) The extent of genetic diversity that is present in the gene pool, (2) The heritability of the characteristic, and (3) Analytical methods for the quantitative measurement of these factors.

The following is a list of some of the restrictions that come with breeding for improved nutritional quality:

1. The interplay of multiple genes, each of which has a relatively minor effect on its own, determines the qualities that make up a product, such as the amount of protein and seed oil it contains. To put it another way, they are controlled by polygenes as well as modifier genes. So, the probabilities of picking good genotypes are smaller than they are for qualities that are controlled by oligogenes. Additionally, environmental variables exert a significant amount of influence on polygenic traits.
2. There is an inverse relationship between seed protein content and grain yield in both cereals and pulses. Grain yield in maize has a negative correlation with both the amount of seed protein and seed

oil it contains. In such a circumstance, it is quite difficult to accomplish simultaneous improvement in yield and protein percentage, as well as yield and oil percentage. In addition, there is a contradictory relationship between seed oil and seed protein. As a result, neither can be improved while the other is.

3. The improvement in quality may result in a smaller grain size and lower yield in some circumstances. In barley, for instance, selecting for a high lysine content results in a decrease in both grain size and grain production.
4. On occasion, the desirable trait can be discovered in wild relatives or species. Transferring a trait from a wild species to a domesticated one can provide a number of challenges, including the possibility of incompatibility between the two species, hybrid inviability, hybrid sterility, and coupling of bad genes with good genes.
5. The work entails a significant amount of chemical investigation. In order to complete the study, biochemists need to work closely together, which might be a constraining aspect at times.

Conclusion

The concept of quality refers to a collection of characteristics rather than just one of them. Because of this, in contrast to yield, there is not a single estimate for it. To determine the quality of the product, it is necessary to first evaluate the characteristics of quality in an isolated fashion and then to interpret the results. The already extensive list of characteristics that a breeder needs to choose for in order to create a variety that is acceptable includes quality traits as well.

References

Axtel, J.D. 1981. Breeding for improved nutritional quality. In K.J. Frey (ed.) plant breeding II, pp.365-432. IOWA State University Press, Ames.

Bhatia, C.R. and Rabson, R. 1976. Bioenergetic considerations in cereal breeding for protein improvement, Science 194: 1418-1420.

Gupta, V.P. 1992. Genetic improvement and management of quality in crop plants. Indian J. Genet. 52: 207-212.

Chopra, V.L. 1989. Plant Breeding. Oxford and IBH publishing Co. Pvt. Ltd., New Delhi.

Singh, D.P. 1991. Genetics and Breeding of Pulse Crops. Kalyani Publishers, New Delhi.

Singh, P. 1998. Cotton Breeding. Kalyani Publishers, New Delhi.

Chapter 12

Genome Editing Approaches to Fulfill Breeder's Dream

Chamin Chimyang[1], Nilutpal Saikia[2], Sweta Kumari[3] and N. Anthony Baite[2]

[1]Ph.D. Scholar, Department of Genetics and Plant Breeding
[2]Ph.D. Scholar, Department of Agronomy
[3]Ph.D. Scholar, Department of Soil Science and Agricultural Chemistry
Banaras Hindu University, Varanasi, Uttar Pradesh
**e-mail: chimyang11@gmail.com*

Introduction

Overview of Plant Breeding and Limitations of Traditional Methods

Plant breeding is a crucial aspect of agriculture as it helps to develop new plant varieties that have improved characteristics. These improved characteristics can include increased yield, enhanced nutritional quality, and improved resistance to pests and diseases. Over the years, plant breeding has evolved, and conventional methods such as hybridization and mutation breeding have been used to create new plant varieties. However, traditional methods of plant breeding have their limitations. One major limitation of traditional methods is that they can be time-consuming. The process of creating a new plant variety through crossbreeding can take several years, and even then, there is no guarantee that the desired traits will be expressed in the offspring. Additionally, traditional methods may not always result in precise changes to the plant genome. For example, cross breeding can result in the introduction of unwanted traits, and mutation breeding can cause unintentional and random changes to the plant genome.

Introduction to Genome Editing as a Tool for Crop Improvement

Genome editing is a relatively new technique that has revolutionized plant breeding by allowing for precise and targeted changes to be made to the plant genome. The development of the CRISPR-Cas9 system has enabled researchers to make specific alterations to the DNA of plants, leading to the introduction of desired traits or the removal of unwanted ones. One of the key advantages of genome editing over traditional methods is that it allows for precise and targeted changes to be made to the plant genome. This precise control over the plant genome enables researchers to introduce specific traits, such as disease resistance or improved nutritional content, with greater accuracy and efficiency. Moreover, genome editing can be done in a much shorter time frame compared to traditional methods, as it does not involve the time-consuming process of crossbreeding. Furthermore, genome editing allows for the creation of new plant varieties that are not found in nature. This opens up new possibilities for improving crop plants, such as the introduction of traits that provide increased resistance to pests, disease, and environmental stress. This increased resilience can lead to improved crop yields and stability, contributing to food security and sustainability.

Overview of Genome Editing Techniques

1. Zinc Finger Nucleases (ZNFs)

Prior to getting into the topic of ZFNs gene editing tool, it is important to understand zinc finger motifs. Zinc finger motifs are nothing but small, functional independently folded domains that requires the co-ordination of one or more zinc (Zn) ions for stabilizing its structural motifs. The Zn ion will be like the heart of this structure from where the other protein structures will be protruding out like a finger.

It was discovered in the year 1989 in transcription factor (TF IIIA) of *Xenopus laevis* (African clawed frog) and from there onwards many studies has been conducted revealing that the zinc finger motifs vary in size and in their functions as well from DNA or RNA binding, membrane association or protein-protein interaction.

The classical structure of these zinc finger motifs is the presence of conserved cysteine and histidine ligands (also known as Cys_2 His_2 zinc fingers). They are the most abundant protein in eukaryotic genome and many functions as transcription factors for recognition of specific DNA sequences. These proteins have highly conserved linker sequence known as TGEKP that connects adjacent finger. Many debates have been already discussed regarding whether the conserved sequence plays an active or passive role in DNA binding. One of the functions of these linker sequences is to cap C-terminus of the preceding helix and this C-capping interaction is not present in the free protein and is induced only when zinc finger binds to DNA. It was found that in free protein the conserved linker sequence are disoriented and adopt the structure with restricted backbone flexibility upon binding to the protein.

It was assumed from a model of DNA binding that the flexibility of the linker will facilitate diffusion along the DNA strand as zinc finger protein searches for its specific binding site and once it has been recognized, structure is then induced in the linkers which will act as "snaplocks", holding the finger in proper orientation for optimal interaction in the major groove.

Now with the above knowledge of what a zinc finger motif is, we can try to depict the picture of how it can be engineered into a gene editing tool. So, basically, Zinc finger Nuclease (ZNFs) is a gene editing tool that hires two domain- a) Cys_2 His_2 ZF domain and b) a nuclease domain of Type IIS restriction enzyme, *Fok*1, which are fused together to give the following gene editing tool. The Cys_2 His_2 ZF domain guides the nuclease domain and cleaves the target DNA sequence to introduce DSB. But since, *Fok*1 cleavage domain is enzymatically active only in the form of a dimer, the ZNFs also has to undergo dimerization. Thereby, we can say that it is designed in such a way that it recognizes the target site in a tail-to-tail conformation and each monomer will be attached to the target site by its half which is separated by a spacer.

In plant genome editing, ZNFs are used to introduce mutations or modifications into the plant genome. The ZNFs bind to the target sequence in the DNA and cut it, leading to a double-stranded break (DSB) in the DNA. The cellular machinery then attempts to repair the DSB, either through non-homologous end joining (NHEJ) or homology-directed repair (HDR). NHEJ is an error-prone repair mechanism that can introduce insertions or deletions (indels) at the break site, leading to disruptions in the target gene and potentially introducing new traits. HDR is a more precise mechanism, where the repair template is provided by exogenously added DNA, and it can lead to the precise insertion, deletion, or substitution of the target sequence.

ZNFs have been used to edit various genes in different plant species, including model plants like Arabidopsis, and crops like rice, soybean, and tomato. Applications of plant genome editing using ZNFs include improving crop yield and stress tolerance, enhancing nutrient content, and reducing the dependence on pesticides and herbicides. For example, ZNFs have been used to introduce resistance to diseases such as rice blast and soybean cyst nematode, and to increase tolerance to abiotic stress such as drought and high salinity.

Advantages of Zinc Finger Nucleases (ZNFs)

- ☆ Precision: ZNFs can precisely target specific DNA sequences, allowing for more controlled and specific gene editing.

- ✰ Versatility: ZNFs can be engineered to target almost any desired DNA sequence, making them highly versatile in their applications.
- ✰ Ease of Use: ZNFs can be delivered into cells using a variety of methods, including viral vectors or electroporation.

Disadvantages of Zinc Finger Nucleases (ZNFs)

- ✰ Complexity: The process of engineering ZNFs can be complex, requiring specialized knowledge and expertise.
- ✰ Cost: ZNF technology is relatively expensive compared to other gene editing techniques.
- ✰ Off-Target Effects: There is a risk of off-target effects, where the ZNFs may inadvertently target and modify other DNA sequences besides the intended target.
- ✰ Immune Response: There is a potential for an immune response to the ZNFs, particularly if they are delivered via viral vectors.

To increase the specificity of ZF domain, the following consideration can be taken

- ✰ To improve ZF domain for DNA binding
- ✰ To optimize the linker sequence connecting the ZF domain to the endonuclease domain
- ✰ To regulate the cleaving of *Fok*1 endonuclease.

2. TALENs

TALENs were inspired by zinc finger motifs and Xanthomonas-derived Transcription Activator Like Effectors (TALE). AvrBs from Xanthomonas campestris pv. vesicatoria was the first TAL effector. And just like ZFNs gene editing tool, TALENs is also constructed via two important individual components. As the name indicates it can be divided into separate components- (a) TALE (Transcription activator-like effector) and (b) Nuclease which has been fused together to construct what now we see as the gene editing tool.

But prior to leaping into the topic, we have to get a good understanding of TALEs. TALEs are used by *Xanthomonas* pathogen which is a host of many domesticated crop that is used to infect plants. They are basically type III-secreted proteins that act as modular, sequence specific transcription factor in the plant cells. The N- and C-terminal provide the function of- (a) allows translocation into plant cells (by N-terminal type III secretion signal), (b) import into the nucleus (by C-terminal nuclear localization signal) and lastly (c) their interaction with the eukaryotic transcriptional machinery (C-terminal acidic activation domain).

TALE also has central repeat domain (CRD) which is a sequence specific DNA binding and consists of nearly identical 33-35 amino acids repeat sequences that will differ primarily at 12 and 13 positions and are known as repeat variable di-residues (RVD). One repeat can bind to 1 DNA base and the specificity is determined by RVD and if there is alteration in repeat number or in RVD order the target specificity will change accordingly.

Thereby, we can say that just like ZFNs, TALENs also employs two components- (a) non specific *Fok*1 nuclease domain and (b) customizable DNA binding protein, *i.e.*, TALE protein. As said earlier, each individual TALE repeat will bind to a single DNA base whose identity is determined by the two hypervariable residues which is of course, mainly primarily present in 12 and 13 position of the repeat array. Therefore, number of repeats corresponds to the length of its target site and thus, building a simple correlation between hypervariable residues and the base bound by each repeats. Additionally, it was also reported that thymine was conserved at the 5′ end just prior to the base bound by the first repeat in the array.

All the constructed TALE repeat arrays reported till date uses 4 domains which consists of hypervariable residues, NN, NI, HD, and NG which for identification of guanine, adenine, cytosine and thymine respectively. But recent studies have also discovered some other hypervariable such as NK, which has been reported to be more specific for guanine but with less activity. Likewise, another study has reported of NH which is more specific than NN but with slight low activity.

The selection of the vector and assembly procedure would vary based on need and inventiveness, but the fabrication of domains with nearly identical repetitions remains a difficulty. A variety of strategies have been proposed for the production of DNA-binding engineered TALE domains. The previous procedures were laborious and time-consuming, but the more recent automated approaches are fast and convenient. For the initiation of desired editing, Platinum TALENs is used in the majority of contemporary investigations. In vitro and in vivo, TALENs have had a significant effect, enabling the modification of sequences in several model species that were previously inaccessible. TALENs not only gave the benefit of easy customization in comparison to Zinc fingers, but also increased the diversity and capacity of gene editing reagents. In addition to nuclease, TAL effector proteins might interact to gene activators and receptors. TALENs were much simpler to design than ZFNs. Identification of RVD on repeat sections of TALE improved DNA binding effectiveness and reduced manufacturing time, opening the path for a library of TALENs capable of targeting all genes in a genome. Additionally, the repetition areas' extensible capacity was a bonus. In general, it was designed for binding 18bp or longer target sequences to obtain a better degree of specificity and affinity than ZFN, which would bind to shorter sequences ranging from 9 to 12 bp. A further advantage of TALENs over ZFN is a greater degree of flexibility for site selection, which increases the likelihood of locating potential sites for each 100bp of a genome. Despite the fact that methylation at the target location created a barrier for the binding affinity of TALENs, methylated DNA bases might be present in RVD code that has been modified. Despite all the benefits provided by TALENs, the larger size requirement of typically 3kb for its construction continues to be its major disadvantage compared to ZFN, which requires a smaller construct of approximately 1kb; since larger size constructs causes several complications with respect to packaging in the vector molecule as well as decreased expression efficiency in the host organism.

Unlike ZFNs, TALENs does not require reconstruction of base linkage between repeats arrays with the ability of targeting single site in a genome but its only limitation is that TALE binding site should start with a "T" base. Nevertheless, the ease of assembling the TALE repeats has reported the constructions of TALENs library with a target of 18,740 human protein-coding genes. Nature Methods dubbed it 2011's "Method of the Year" in recognition of this tremendous advancement in the field of genetic engineering.

Advantages of TALENs (Transcription Activator-Like Effector Nucleases)

- ✰ High specificity for target DNA sequences, reducing the risk of off-target effects.
- ✰ Ability to target a wide range of genomic sequences, making TALENs useful for a variety of genome editing applications.
- ✰ Flexibility in design, allowing researchers to create TALENs that specifically target any desired DNA sequence.

Disadvantages of TALENs

- ✰ Complex design process, which can be challenging and time-consuming.
- ✰ Difficult to scale up the technology for large genome editing projects.
- ✰ May not work effectively in all cell types, limiting the range of applications for TALENs.
- ✰ High cost associated with the design and production of TALENs.

3. CRISPR

The CRISPR-Cas9 system is a revolutionary genome editing technique that allows for precise, targeted modification of DNA. This technology is based on the natural defense mechanism of bacteria against invading viruses, called CRISPR (Clustered Regularly Interspaced Short Palindromic Repeats).

Mechanism of Action

1. The CRISPR-Cas9 system consists of two key components: the RNA-guided endonuclease Cas9 and a guide RNA (gRNA) molecule. The

gRNA molecule is composed of a CRISPR RNA (crRNA) and a trans-activating RNA (tracrRNA).

2. The crRNA component of the gRNA molecule is designed to recognize a specific target DNA sequence, while the tracrRNA component helps to form the active complex with Cas9.
3. The gRNA molecule guides Cas9 to the target DNA sequence, where the nuclease activity of Cas9 cleaves both strands of the DNA.
4. This DNA double-stranded break triggers the cell's natural DNA repair mechanisms, which can lead to either insertion or deletion of specific genetic sequences. This allows researchers to add, remove, or replace specific DNA sequences in a targeted manner.

The CRISPR-Cas system is a way for single-celled organisms like archaea and bacteria to fight off viruses. It's made up of spacers from viruses separated by repeated palindromic sequences. The system works in three stages: adaptation, expression, and interference. In the adaptation stage, the spacers from the virus are added to the organism's memory. Then, a complex of Cas proteins recognizes a PAM motif on the target DNA and cleaves it. In the expression stage, the CRISPR gene is turned into RNA and processed. In the interference stage, the crRNA, with the help of Cas proteins, recognizes the invading genome and cleaves it. The Cas proteins can either be part of the crRNA complex or be recruited during interference.

The CRISPR-Cas system exhibits a high level of sequence diversity due to ongoing battles with viruses which leads to rapid evolution. This complexity has made it difficult to classify the system. However, Marakova *et al.* (2015) identified 5 types and 16 sub-types of CRISPR-Cas, and divided them into two broad classes. Class 1 consists of multiple Cas proteins that form the crRNA binding complexes for pre-crRNA processing and interference. Class 2 has only one multi-domain crRNA binding protein that performs all the necessary functions for interference and some pre-crRNA processing. The story of engineering the CRISPR system started in 1987 when Yoshizumi Ishino discovered a repetitive sequence while studying a gene in E. coli. In 1989, Francisco Mojica observed similar sequences in a different microbe and concluded that their presence in different organisms indicated an important function. By 2000, Mojica found these sequences in over 20 microbes and Jansen *et al.*, named the system CRISPR. In 2005, Mojica's work showed that CRISPR likely functions as a bacterial immune system. Other researchers also made significant contributions to the field, such as Pourcel *et al.*, who found that CRISPR spacers were acquired from viruses. Bolotin *et al.*, was the first to propose a mechanism for CRISPR immunity based on anti-sense RNA inhibition, but this was later proven incorrect.

In 2011, Phillipe Horvath and colleagues revealed that bacterial clones chosen for phage resistance acquired CRISPR spacers relative to uninfected bacterial clones. There was a substantial correlation between the number of spacers present and the level of resistance shown by the strain. In addition, they analysed the function of the two Cas proteins, Cas7 and Cas9, and discovered that Cas7 was not involved in the mechanism of resistance, as strains carrying virus-derived spacers remained resistant to viral infection despite the deletion of Cas7, indicating that it was not involved in the acquisition of resistance. In the meanwhile, the Cas9 sequence had two nuclease domains (HNH and RuvC) necessary for the resistance mechanism. Viruses with a single base modification in their photospacer sequences were able to bypass CRISPR bacterial resistance, according to another conclusion of the study. Jon van der Oost, Eugene Koonin, and colleagues originally reported the construction of the CRISPR system when they transferred the CRISPR system of one E. coli strain into another that lacked its own CRISPR system for biochemical analysis. The examined CRISPR-Cas system carried out the activities of Cas9 employing Cas3 nuclease and a number of additional proteins that comprised the cascade complex. They knocked out each component and demonstrated that the cascade complex components were involved in the CRISPR RNA processing. In addition, they synthesised spacer sequences from four key genes of lambda phage and proved that strains produced to include these spacer sequences were resistant to lambda phage. Researchers developed a bacterial influenza vaccine. They created anti-sense (complementary to the coding DNA and the mRNA) and sense CRISPR arrays (complimentary to the other strand of DNA). The sense version of CRISPR had far greater effects, showing that the CRISPR system operates on DNA and not RNA; hence, this was offered as a hypothesis. Other researchers, however, disagreed with this theory and were sure that the CRISPR system targeted DNA since RNAi could not prevent the rapid proliferation of viral particles upon infection.

Luciano Marrafani hypothesised that it targeted DNA like restriction enzymes. During his study at Northwestern University, he uncovered a spacer in a strain of Staphylococcus epidermis that resembled a nickase enzyme on a plasmid in S. aureus. He showed that plasmids could only transfer to S. epidermis after disrupting the spacer or nickase sequence, thereby proving that CRISPR targets DNA. Later, Murrafani and Sontheimer introduced a self-splicing intron in the plasmid's nickase gene to show that the CRISPR system targeted DNA, not mRNA. Since they had successfully engineered this system, they concluded that it was basically a restriction enzyme that was extremely programmable.

CRISPR system created DSB was unknown at first. Slyvian Moineau and his team proceeded to study CRISPR and found that certain strains of S. thermophilus were partially resistant to plasmid transformation and one had linearized plasmids. The linearized plasmid was sequenced to create blunt ends by cleaving the strand three nucleotides upstream of the Protospacer Adjacent Motif (PAM). Invading viral DNA is likewise cleaved at this same point relative to the PAM, and the number of spacers matched the number of cuts. The crRNA guides the CRISPR system to split DNA precisely. The CRISPR system was lacking an element. Emmanuelle Charpentier and Jörg Vogel found it. Vogel focused on Next Generation Sequencing microbial genome catalogues whereas Charpentier studied inter-genic regulatory RNA sequences. They sequenced Streptococcus pyrogenes' genome together. They discovered an unique short RNA generated from a region near the CRISPR locus that had near-perfect sequence comparable to the CRISPR repeats and was the third most prevalent class of RNAs after r-RNA and t-RNA. Since this new RNA had a similar sequence to CRISPR repeats, its product must be hybridising to the precursor crRNA to convert it into a mature crRNA. Gene deletion tests verified their hypothesis that this unique tracrRNA was essential for crRNA processing and CRISPR activity. Further research showed that tracrRNA was essential to the Cas9 nuclease complex that cleaves DNA and processes crRNA.

Advantages of CRISPR in Plants

- Precision: CRISPR allows for very precise changes to be made to specific genes or regions of the plant genome.
- Efficiency: CRISPR is faster and more efficient than traditional methods of plant genetic engineering.
- Versatility: CRISPR can be used to perform various types of genome editing, including insertion, deletion, and replacement of DNA sequences.
- Cost-effectiveness: CRISPR is less expensive than traditional methods, making it accessible to more researchers and organizations.
- Speed: CRISPR is a faster method than traditional methods, allowing for rapid progress in plant genetics research.

Disadvantages of CRISPR in Plants

- Off-target effects: There is a risk of unintended changes to other parts of the genome, which could lead to unintended consequences.
- Complexity: CRISPR is a complex technology that requires specialized equipment and trained personnel to perform.
- Regulatory issues: There are still regulatory and safety concerns about the use of CRISPR in plants, particularly for crops intended for food and animal feed.
- Gene silencing: In some cases, CRISPR can lead to unintended gene silencing, which could impact the plant's growth and development.
- Inheritance of edited traits: There are still questions about the inheritance and stability of edited traits in plants, particularly in crops.

Applications of Genome Editing in Crop Improvement

Genome editing tools such as CRISPR-Cas9 have the potential to revolutionize crop improvement by enabling precise, targeted modification of crop genomes. In this book chapter, we will explore the various applications of genome editing in crop improvement and how these tools can be used to increase crop yield, improve disease and stress tolerance, and enhance nutritional quality.

Increasing Crop Yield

1. Improving Photosynthesis Efficiency

Photosynthesis is the process by which plants convert sunlight into energy, and it is a major determinant of crop yield. Genome editing tools can be used to modify genes involved in photosynthesis, increasing the efficiency of this process and leading to higher crop yields. For example, researchers have used genome editing to modify genes involved in chlorophyll biosynthesis, which has resulted in increased chlorophyll content and improved photosynthesis in crop plants.

2. Enhancing Nutrient Uptake and Utilization

Another important factor affecting crop yield is nutrient uptake and utilization. Genome editing tools can be used to modify genes involved in nutrient uptake and transport, leading to enhanced utilization of essential nutrients and increased crop yields. For example, researchers have used genome editing to modify the expression of genes involved in the uptake of nitrogen and phosphorus, resulting in improved growth and increased yields in crop plants.

Improving Disease and Stress Tolerance

1. Resistance to Biotic and Abiotic Stressors

Diseases and environmental stressors such as drought and high temperatures can significantly reduce crop yields. Genome editing tools can be used to introduce resistance to these stressors by modifying genes involved in disease resistance and stress tolerance. For example, researchers have used genome editing to introduce resistance to viral diseases in crops, resulting in reduced disease incidence and improved yields.

Another example is the use of the CRISPR system to make cassava resistant to brown streak disease and mosaic virus, spinach resistance to downy mildew, apple resistance to fire blight, orange resistance to greening, maize resistance to lethal necrosis, grapevine resistance to powdery mildew, and tomato resistance to Yellow Leaf Curl Virus. While TALENs has been used to increase wheat's resistance to fungi and almonds' resilience to worms, it has also been used to wheat's resistance to nematodes. ODM has been used to confer Phytophthora resistance to potatoes. Rapid Trait Development System (RTDS), the proprietary gene editing breeding platform of Cibus, was utilised to create herbicide-resistant plants. Canola with herbicide resistance is already on the market in Canada.

2. Enhancing Tolerance to Environmental Conditions

In addition to disease resistance, genome editing tools can also be used to enhance the tolerance of crops to environmental conditions such as high temperatures, drought, and salinity. For example, researchers have used genome editing to modify genes involved in water utilization and stress tolerance, leading to improved growth and increased yields in crops under harsh environmental conditions.

Improving Nutritional Quality

1. Introduction of Essential Micronutrients

Crop plants are often deficient in essential micronutrients such as iron, zinc, and vitamin A. Genome editing tools can be used to introduce these micronutrients into crops, improving their nutritional quality and providing important health benefits for consumers. For example, researchers have used genome editing to introduce iron and zinc into rice, resulting in a rice variety with increased levels of these essential micronutrients.

2. Modifying Composition of Macronutrients

In addition to introducing micronutrients, genome editing tools can also be used to modify the composition of macronutrients such as carbohydrates, proteins, and fats. For example, researchers have used genome editing to modify the starch content and composition in crops such as potato, leading to a higher quality product with improved nutritional value.

The J.R. Simplot Company, located in the United States, created non-browning potatoes by silencing the asparagine synthetase-1 gene, which decreased acrylamide generation by 58-72 per cent when potatoes were baked, fried, or roasted at high temperatures. Since asparagine is a precursor to acrylamide, a carcinogen, reducing asparagine synthesis would eventually reduce acrylamide levels. Calyxt (USA), a plant-based synthetic biotechnology firm, employed TALENs to generate high fibre content in wheat and high oleic oil content in soybean, whereas Yield10 Bioscience, an agricultural bioscience company in the USA, developed Camelina sativa with increased omega-3 oil.

DuPont Pioneer, the first business to licence CRISPR technology, has stated on April 18, 2016, that its first CRISPR-edited crop would be a waxy corn. Usually, a normal corn kernel includes 75 per cent amylopectin and 25 per cent amylose, however a deletion in the waxy gene results in waxy corn kernel with over 97 per cent amylopectin, which basically removes amylose from the kernel. The starch extracted from these kernels will have novel applications in the culinary and industrial sectors.

Challenges and Considerations in the Use of Genome Editing in Crop Improvement

In several sectors, including agricultural sciences, genome editing technology has sparked a revolution. However, with its recent advancements comes a larger responsibility on how far this technology may be used. The laws and regulations governing genome-edited plants are developing swiftly to accommodate new technology.

1. Regulatory Landscape

The regulation of genome editing in crops is a complex issue that varies greatly between countries. In some countries, the regulation of genetically modified crops is relatively lenient, while in others it is much more strict.

- National and international regulations: The national and international regulations for genome editing in crops can impact the development and commercialization of these crops. For example, some countries have regulations in place that restrict the use of genetically modified crops, while others have regulations that allow their use but impose specific requirements and conditions. In 2018, the EU Court of Justice (ECJ) declared that all genome-edited organisms must be classified as GMOs in Europe. In addition to the EU, New Zealand oversees genome editing via its GM biosafety standards and court decisions. This has a severe effect on agricultural developments, and as a result, several major firms have relocated their sophisticated breeding initiatives elsewhere. Only 8 per cent of agricultural patents originate in Europe, whereas 60 per cent originate in China and 26 per cent in the United States. In January of 2020, India produced a draught paper titled "Regulatory Framework and Guidelines for Risk Assessment" about genome-edited creatures. The paper recommended a tiered system for categorising group products depending on risk, with simple or a few base-pair alterations recognised as low risk – "Group 1" – and insertion of big or alien DNA designated as high danger – "Group 3." The proposal addresses genome-edited plants, animals, and human cells in a single document, which is uncommon among biosafety regulations, and typically conflates the danger associated with each. Therefore, India became an illustration of the significance of deregulation. In 2015, Argentina became the first nation to announce that crops would not be subject to biosafety regulations if they do not contain any foreign DNA. Chile, Brazil, and Colombia followed shortly after.
- Approval process for genetically modified crops: The approval process for genetically modified crops is also a factor that varies between countries. Some countries have a streamlined approval process that allows for the rapid commercialization of genome-edited crops, while others have a more complicated and time-consuming process. This

can greatly impact the development and commercialization of these crops, and could potentially limit their potential benefits. In 2018, the United States Department of Agriculture (USDA) opted not to regulate new genome editing-based breeding methods. If they do not contain novel DNA, Japan does not control plant varieties differently from traditionally grown kinds. China has failed to create a regulatory framework for the assessment of genome-edited plant products, despite having the most papers and patents for gene-editing uses in agriculture.

2.. Potential Unintended Consequences

The use of genome editing in crops also poses potential unintended consequences that must be taken into account.

1. Gene flow and the potential impact on non-target species: One of the main potential unintended consequences of using genome editing in crops is gene flow. This occurs when genes from one species are transferred to another species through cross-pollination. This can lead to unintended impacts on non-target species and could potentially cause ecological imbalances. For example, if a gene from a genetically modified crop is transferred to a wild relative, it could result in the creation of a new invasive species.
2. Ethical and social considerations: The use of genome editing in crops also raises ethical and social considerations. For example, some people may question the potential impact of these crops on biodiversity, and whether the development and control of these crops should be in the hands of corporations or government organizations. There are also concerns about the potential impact of genetically modified crops on food security and food sovereignty, as well as on farmers who may be impacted by the commercialization of these crops.

Conclusion

Summary of the Potential of Genome Editing for Crop Improvement

Genome editing is a breakthrough technology that allows scientists to precisely and efficiently modify the genetic code of an organism. In the field of agriculture, this technology has the potential to transform the way crops are grown and improve food security by addressing some of the biggest challenges facing farmers today.

The CRISPR-Cas9 system has been used to modify the genomes of numerous crop species, including rice, wheat, maize, and soybeans. For example, researchers have used CRISPR-Cas9 to produce rice plants that are more resistant to pests and disease, which could have a major impact on food security in developing countries. Additionally, researchers have used this technology to produce crops that are more drought-tolerant and require fewer inputs, such as water and fertilizer, making them more environmentally sustainable.

Another promising application of genome editing is the production of crops that are more nutritious. For example, researchers have used CRISPR-Cas9 to produce rice plants that are rich in beta-carotene, a precursor of vitamin A. This could have a major impact on malnutrition in developing countries, where vitamin A deficiency is a major public health problem.

Future Perspectives and Ongoing Research in the Field

In terms of future perspectives and ongoing research, there is a growing body of research that is exploring the use of genome editing for crop improvement. For example, researchers are working on developing new and more efficient genome editing tools, such as RNA-guided nucleases that can make multiple changes to the genome simultaneously. Additionally, there is a growing interest in using genome editing to produce crops that are more sustainable and environmentally friendly, such as crops that require fewer inputs and emit fewer greenhouse gases.

However, it is important to consider the ethical and social implications of this technology, as well as to ensure that it is used in a responsible and sustainable manner. For example, there are concerns about the potential for unintended consequences of modifying the genomes of crops, such as the creation of new plant diseases or the evolution of pests that are resistant to the new crop varieties.

REFERENCES

Carroll, D. (2021). A short, idiosyncratic history of genome editing. *Gene and Genome Editing, 1*, 100002.

Cathomen, T., and Joung, J. K. (2008). Zinc-finger nucleases: the next generation emerges. *Molecular Therapy, 16*(7), 1200-1207.

Gaj, T., Gersbach, C. A., and Barbas, C. F. (2013). ZFN, TALEN, and CRISPR/Cas-based methods for genome engineering. *Trends in biotechnology, 31*(7), 397-405.

Joung, J. K., and Sander, J. D. (2013). TALENs: a widely applicable technology for targeted genome editing. *Nature reviews Molecular cell biology, 14*(1), 49-55.

Laity, J. H., Lee, B. M., and Wright, P. E. (2001). Zinc finger proteins: new insights into structural and functional diversity. *Current opinion in structural biology, 11*(1), 39-46.

Randhawa, S., and Sengar, S. (2021). The evolution and history of gene editing technologies. *Progress in Molecular Biology and Translational Science, 178*, 1-62.

Ricroch, A. (2019, August). Global developments of genome editing in agriculture. In *Transgenic research* (Vol. 28, pp. 45-52). Springer International Publishing.

Schandry, N., Jacobs, J. M., Szurek, B., and Perez Quintero, A. L. (2018). A cautionary TALE: how plant breeding may have favoured expanded TALE repertoires in Xanthomonas. *Molecular plant pathology, 19*(6), 1297.

Schmidt, S. M., Belisle, M., and Frommer, W. B. (2020). The evolving landscape around genome editing in agriculture: Many countries have exempted or move to exempt forms of genome editing from GMO regulation of crop plants. *EMBO reports, 21*(6), e50680.

Chapter 13

Special Breeding Methods: Marker-Assisted Crop Improvement

S. Bhaskar Reddy[1]*, *S. Anand*[1], *M.O. Ankitha*[1] *and B. Arani Jyothi*[2]

[1]*Ph.D. Scholar, Department of Genetics and Plant Breeding, College of Agriculture, Vellayani, Thiruvananthapuram, Kerala*
[2]*Ph.D. Scholar, Department of Genetics and Plant Breeding, S.V. Agriculture College, Tirupati, Andhra Pradesh*
e-mail: bhaskarreddy1996@gmail.com

Introduction

The expanding world population necessitates the production of additional food grains. The area under cultivation, in particular, is decreasing due to urbanization and the building of new roads and structures. Therefore, the future of plant breeding should strongly emphasize yield, stability, and sustainability of crops. The primary characteristics of focus are durable disease resistance, abiotic stress tolerance, and nutrition and water use efficiency. Many of these qualities are not simple to choose phenotypically, if they have been identified and mapped as genes or QTLs that can be chosen based on marker genotypes. DNA markers improve the efficiency and precision of conventional plant breeding via marker-assisted selection (MAS).

Molecular Markers

Genetic information in the form of DNA is stored in cell at three places, *viz.*, nucleus, chloroplast and mitochondria. The genetic variation arises in this stored information due to the change in the DNA sequence by base substitution, deletion, insertion, duplication and inversion. These variations are manifested as variation in trait expression (at phenotype level) or amino acid sequence variation (at protein level) or nucleotide sequence variation (at DNA level). In general, marker is anything that marks or tags or identify an individual. Any heritable variation following Mendelian inheritance can be called as genetic marker. Genetic markers facilitate mapping and discovery of valuable genes hence, they are widely used in crop improvement programmes. Compared to conventional phenotypic selection, marker assisted selection (MAS) provides more reliability, efficiency and rapidity in plant breeding programmes. Genetic markers are of three types, *viz.*, morphological markers, biochemical markers and DNA based markers.

- ☆ **Morphological markers**: These are related to shape, size, colour and surface of various plant parts.
- ☆ **Biochemical marker**: Such markers are related to variation in protein and amino acid banding patterns.
- ☆ **DNA markers**: Typically, small regions of DNA showing sequence polymorphism in different individuals with in a species. *E.g.*, RAPD, AFLP, SSR,CAPS *etc.*

Classification of Molecular Markers

Based on banding pattern molecular markers can be characterized as a single locus and multi locus. Based on detection method markers can be characterised as hybridization based and PCR based.

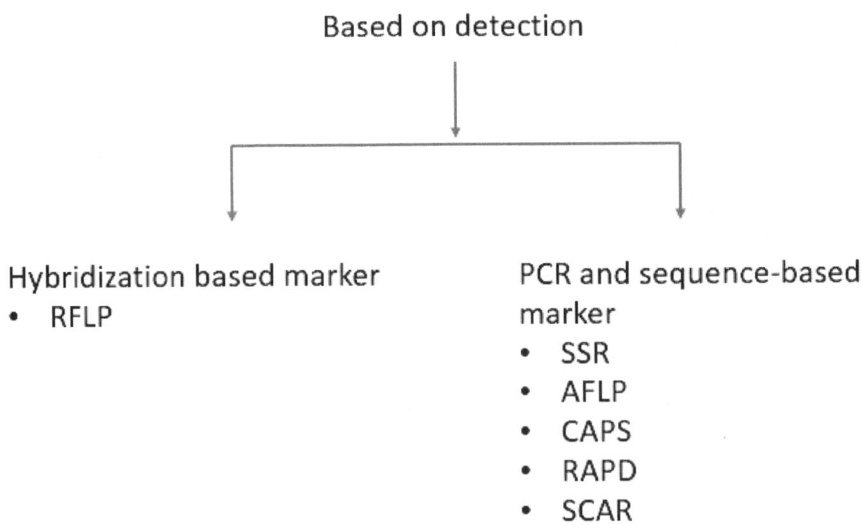

Figure 13.1: Classification of Molecular Markers Based on Detection Technique.

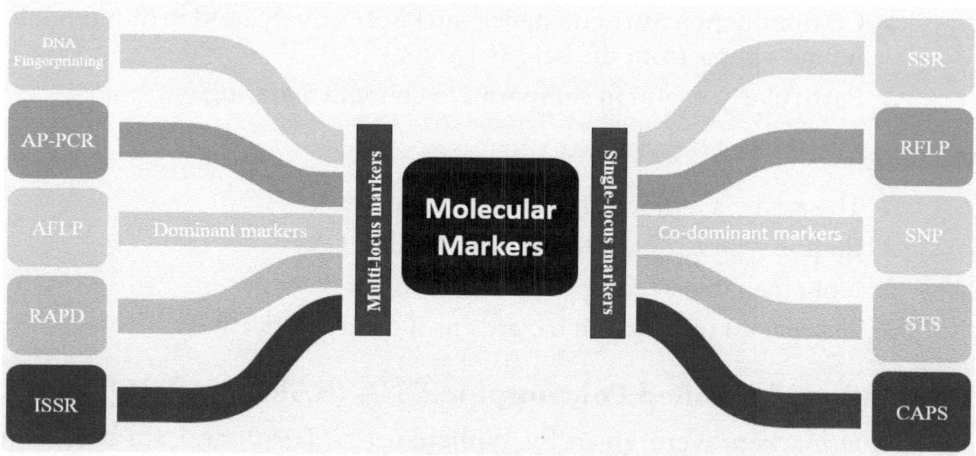

Figure 13.2: Classification of Markers Based on Locus Representation.

Characteristics of Ideal Genetic Marker

- ☆ High level of polymorphism
- ☆ Stability, no influence of environment
- ☆ Simplicity of observation
- ☆ Non-epistatic
- ☆ Co-dominancy
- ☆ Genome-wide distribution
- ☆ Low cost
- ☆ Reproducibility
- ☆ Portability between species
- ☆ Mendelian inheritance

Types of DNA Markers

1. Restriction Fragment Length Polymorphism (RFLP)

RFLP is hybridization -based marker, developed by Botstein *et al.* (1980). This is the polymorphism which is detected based on the differential hybridization of cloned DNA to DNA fragments in a sample of restriction enzyme digested DNAs. Hence, the polymorphism detected in the fragment length is due to variation in restriction site(s) of the enzyme used and these RFLPs are defined by specific enzyme-probe combinations.

Strength

- ☆ Robust and reproducible

- ☆ Co-dominant nature and hence can be effectively used to differentiate homozygotes from the heterozygotes
- ☆ Particularly useful in comparative genome mapping

Weakness

- ☆ The assay is tedious and time consuming
- ☆ Requires large quantities of high quality DNA
- ☆ Not amenable to complete automation
- ☆ Associated with health hazards as it used radioactivity

2. Random Amplified Polymorphic DNA (RAPD)

RAPD markers were given by Williams *et al.* (1990) and are based on the differential PCR amplification of a sample of DNAs from random oligonucleotide sequence. RAPDs are genetically dominant in nature. RAPD uses single random primer usually of 10 nucleotides long, to amplify inverted repeats in the genome and can anneal at multitude of genomic locations.

Strength

- ☆ Fast and so called Rapid markers
- ☆ Requires less quantities of DNA
- ☆ Needs limited investment in time and training
- ☆ Commercially available
- ☆ No radioactivity involved
- ☆ No need of sequence information

Weakness

- ☆ Lack of reproducibility in markers patterns across lab and experiments
- ☆ Dominant in nature

3. Amplified Fragment Length Polymorphism (AFLP)

AFLP markers were given by Vos *et al.* (1995) and these polymorphism are generated using a procedure that combines restriction digestion and PCR amplification. The basic procedure is as follows:

- ☆ Digestion of genomic DNA with a combination of two restriction enzyme- a rare and a frequent cutter, *e.g.* EcoRI and MSeI; EcoRI and PStI.
- ☆ Ligation of double stranded adapters ot cut ends of DNA fragments
- ☆ Pre-selective amplification
- ☆ Selective amplification

- ☆ Separation of amplified fragments
- ☆ Visualization using autoradiography

Strength

- ☆ Stable amplification and high repeatability
- ☆ Can generate fingerprints of any DNA regardless of their origin, so highly suited for DNA fingerprinting
- ☆ Can act as bridge between genomic and physical maps
- ☆ Hyper-variability

Weakness

- ☆ Time consuming procedure
- ☆ Required significant technical skills and financial resources
- ☆ Dominant in nature

4. Simple Sequence Repeats (SSR)

These are also called as microsatellites or Short Tandem Repeats (STRs) or Simple Sequence Length Polymorphisms (SSLPs). These are variable tandem repeats ranging from 1-10 nucleotides, dispersed throughout the genomes of most eukaryotic organisms. These repeats may be di, tri, or tetra-repeats with repeat pattern being perfect, imperfect, compound or complex. If the repeat number is more than 10 nucleotides, they are referred as minisatelites. The conserved regions flanking these repeats are used to design the primers for PCR based amplification. The polymorphism discovered is due to variation in number of repeats in the individuals.

Strength

- ☆ Abundant and uniformly distributed in the genome
- ☆ Hypervariable (large number of allele per locus)
- ☆ Co-dominant markers
- ☆ Highly reliable and reproducible assay
- ☆ Marker systems of choice in most of crop improvement programmes

Weakness

- ☆ High developmental cost
- ☆ Often primers are species specific.
- ☆ Needs prior information of the sequence to design primers

5. Single Nucleotide Polymorphisms (SNPs)

The differences which are found at single nucleotide position are referred to as single nucleotide polymorphism or SNPs. This type of polymorphism results due to substitution, deletion or insertions. This is mostly biallelic and co-dominant in nature.

Strength

- ✰ Abundant
- ✰ Sequence based robust markers
- ✰ Fast and high throughput
- ✰ Genome-wide distribution
- ✰ Possible to discovery of new alleles, hence useful in allele mining
- ✰ Co-dominant

Weakness

- ✰ Most SNPs are biallelic and thus less informative than SSRs
- ✰ SNP assay is costly
- ✰ Sophisticated equipment and considerable efforts required

Each marker system has its own advantages and limitations. They vary with each other with respect to genomic abundance, locus specificity, polymorphism level, reproducibility, technical requirements and financial investments. Hence, due care has to taken before choosing marker system for different genetic studies.

Applications of Molecular Markers

- ✰ Genetic mapping of traits
- ✰ Marker assisted selection
- ✰ Protection of genotypes
- ✰ Hybrid purity testing
- ✰ Diversity analysis
- ✰ Marker assisted gene pyramiding

While there are several applications of DNA markers in breeding, the most promising for cultivar development is called marker-assisted selection (MAS).

Marker-Assisted Selection

Marker assisted selection (MAS) is an indirect selection process where a trait of interest is selected based on a marker (morphological, biochemical or DNA/RNA variation) linked to a trait of interest. For example, using MAS to select individuals with disease resistance involves identifying a marker allele that is linked with disease resistance rather than the level of disease resistance. The assumption is that the marker associates at high frequency with the gene or quantitative trait locus (QTL) of interest, due to genetic linkage (close proximity, on the chromosome, of the marker locus and the disease resistance-determining locus). MAS can be useful to select for traits that are difficult or expensive to measure, exhibit low heritability and/or are expressed late in development. The term 'marker-assisted selection' was first used by Beckmann and Soller in 1986. However, the first markers that allowed indirect selection of a trait of interest were morphological markers. In 1923, Sax first reported association of a simply inherited genetic marker with a quantitative trait in plants when he observed segregation of seed size associated with segregation for a seed coat color marker in beans (*Phaseolus vulgaris* L.). In 1935, Rasmusson demonstrated linkage of flowering time (a quantitative trait) in peas with a simply inherited gene for flower colour.

Molecular markers are not normal genes as they usually do not have any biological effect. Instead, these are the constant landmarks in the genome. They are identifiable DNA sequences, found at specific locations of the genome, and transmitted by the standard laws of inheritance from one generation to the next generation. They rely on a DNA assay, in contrast to morphological markers that are based on visible traits, and biochemical markers that are based on proteins produced by genes. The molecular marker systems described above allow high-density DNA marker maps (*i.e.* with many markers of known location, interspersed at relatively short intervals throughout the genome) to be constructed for a range of economically important agricultural species, thus providing the framework needed for eventual applications of MAS. Using the marker map, putative genes affecting traits of interest can then be detected by testing for statistical associations between marker variants and any trait of interest. These traits might be genetically simple e.g many traits for disease resistance in plants are controlled by one or a few genes (Young, 1999). Alternatively, they could be genetically complex quantitative traits, involving many genes (*i.e.* so-called quantitative trait loci [QTL]) and environmental effects. Most economically important agronomic traits tend to fall into this latter category.

The goal of plant breeding is to assemble more desirable combinations of genes in new varieties. Selection of desirable plants in early generations for traits of higher heritability is easy but, for traits of low heritability, selection is often postponed until the lines become more homozygous in later generations (F5 or F6). Homozygous breeding lines can be evaluated in replicated field trials. The entire process involves considerable time (5–10 years for elite lines to be identified) and expense. The size and composition of a plant populations, larger the number of genes segregating in a populations to identify specific gene combinations are important aspects and make the breeding process more complex. However, the advent of the new biotechnological tools particularly MAS made the process easy and help in accomplishing the breeding targets quickly.

MAS Schemes in Plant Breeding

These include

1. Early generation MAS
2. Marker-assisted backcross breeding
3. Marker-assisted recurrent section
4. Genome-wide selection using markers
5. Genetic purity test
6. Gene pyramiding

1. *Early Generation MAS*

One of the most intuitive stages to use markers to select plants is at an early generation (especially F2 or F3). The main advantage is that many plants with unwanted gene combinations, especially those that lack essential disease resistance traits and plant height, can be simply discarded. This has important consequences in the later stages of the breeding program because the evaluation for other traits can be more efficiently and cheaply designed for fewer breeding lines (especially in terms of field space).

2. *Marker-Assisted Backcross Breeding*

MAS is very useful in backcross breeding programs, wherein multiple genes/QTLs are involved, particularly in those cases where some of the genes are recessively inherited. There are three levels of selection in which markers may be applied in backcross breeding. In the first level, markers may be used to screen for the target trait, which may be useful for traits that have laborious phenotypic screening procedures or recessive alleles as explained earlier. The second level of selection involves selecting backcross progeny with the target gene and tightly-linked flanking markers in order to minimize linkage drag. This is referred to as 'recombinant selection'. The third level of MAB involves selecting backcross progeny (that have already been selected for the target trait) with 'background' markers. In other words, markers can be used to select against the donor genome, which may accelerate the recovery of the recurrent parent genome. With conventional backcrossing, it takes a minimum of five to six generations to recover the recurrent parent.

Foreground Selection

In foreground selection, the selection is made only for the marker allele(s) of donor parent at the target locus to maintain the target locus in heterozygous state until the final backcrossing is completed. Then the selected plants are selfed and the progeny plants with homozygous DP allele(s) of selected markers are harvested for further evaluation and release. As described above, this is the general procedure of MABC. The effectiveness of foreground selection depends on the number of genes/loci involved in the selection, the marker-gene/QTL association or linkage distance and the undesirable linkage to the target gene/QTL.

Background Selection

In background selection, the selection is made for the marker alleles of recurrent parent in all genomic regions of desirable traits except the target locus, or selection against the undesirable genome of donor parent. The objective is to hasten the restoration of the RP genome and eliminate undesirable genes introduced from the DP. The progress in recovery of the RP genome depends on the number of markers used in background selection. The more markers evenly located on all the chromosomes are selected for the RP alleles, the faster recovery of the RP genome will be achieved but larger population size and more genotyping will be required as well. In addition, the linkage drag also can be efficiently addressed by background selection using DNA markers, although it is difficult to overcome in a traditional backcrossing program.

- Foreground selection and background selection are two respective aspects of MABC with different foci of selection. In practice, however, both foreground and background selection are usually conducted in the same program, either simultaneously or successively.
- In many cases, they can be performed alternatively even in the same generation. The individuals that have the desired marker alleles for target trait are selected first (foreground selection). Then the selected individuals are screened for other marker alleles again for the RP genome (background selection).
- It is understandable to do so because selection of the target gene/QTL is the essential and only critical point for backcrossing program, and the individuals that do not have the allele of target gene will be discarded and thus it is not necessary to genotype them for other traits.

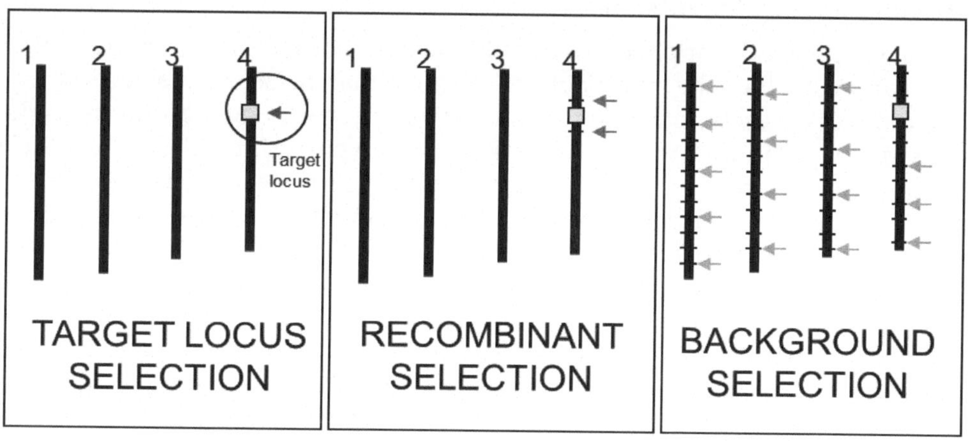

The efficiency of MABC depends upon several factors, such as the population size for each generation of backcrossing, marker-gene association or the distance of markers from the target locus, number of markers used for target trait and RP background, and undesirable linkage drag.

3. *Marker-Assisted Recurrent Selection (MARS)*

The improvement of complex traits via phenotypic recurrent selection is generally possible, but the long selection cycles impose restrictions on the practicability of this breeding method. With the use of markers, recurrent selection can be accelerated considerably. In continuous nursery programs preflowering genotypic information is used for marker-assisted selection and controlled pollination. Thus, several selection-cycles are possible within one year, accumulating favourable alleles in the breeding population. Additionally, it is possible today to define an ideal genotype as a pattern of QTLs, all QTLs carrying favorable alleles from various parents. If individuals are crossed based on their molecular marker genotypes, it might be possible to get close to the ideal genotype after several successive generations of crossings. It is likely that through such a MARS breeding scheme higher genetic gain will be achieved than through marker-assisted backcrossing.

4. *Genome-wide Selection using Markers*

Traditional plant breeding programs rely mainly on phenotypes being evaluated in several environments; selection and recombination are based solely on the resulting data plus pedigree information, when available. Marker assisted selection (MAS) uses molecular markers in linkage disequilibrium (LD) with QTL. Genomic selection (GS) is a new approach for improving quantitative traits in large plant breeding populations that uses molecular markers spanning the whole genome (high density markers and high-throughput genotyping). Genomic prediction combines marker data with phenotypic and pedigree data (when available) in an attempt to increase the accuracy of the prediction of breeding and genotypic values.

In practice, GS is applied in a population that is different from the reference population in which the marker effects were estimated. Genomic selection uses two types of datasets: a training set and a validation set. The training set is the reference population in which the marker effects were estimated; it contains: (1) phenotypic information from relevant breeding germplasm evaluated over a range of environmental conditions; (2) molecular marker scores; and (3) pedigree information or kinship. Hence, marker effects are estimated based on the training set using certain statistical methods to incorporate this information; the genomic breeding value or genetic values of new genotypes are predicted based only on the marker effect. The validation set contains the selection candidates (derived from the reference population) that have been genotyped (but not phenotyped) and selected based on marker effects estimated in the training set.

For quantitative traits, selection based on marker effects alone has dramatically changed standard practices used in plant and animal breeding. However, in public crop breeding programs, the benefits of GS have been studied only through computer simulation. Since marker technology is continuously reducing the cost per data point and increasing the number of available markers, genotyping is currently less costly than phenotyping in an applied plant breeding program. Besides accelerating the selection cycles, GS offers the opportunity to increase the selection gains per unit of time. Therefore, it can be predicted alternating progeny field testing with selection based only on markers should increase the genetic gains per unit.

5. *Genetic Purity Tests*

Molecular markers are routinely used in testing of the released crop varieties. After the completion of the cultivar development its DNA fingerprints are prepared and submitted to the Central Variety Release Committee for notification. These help in maintaining genetic purity as well as resolving the disputes if any later stages.

6. *Gene Pyramiding*

In many cases, the aim of the breeder is either to stack multiple genes for the same trait or combine multiple traits in one final breeding product. The best approach is to perform all those conversions into the same background individually in parallel and then to intercross the final single conversions to combine the traits together. In that case, only MAS is needed at the end since the narrowing of the introgressed regions through foreground selection and the recovery of the recurrent parent through background selection have already been done for each individual trait. The most frequent strategy of pyramiding is combining multiple resistance genes. Different resistance genes can be combined in order to develop broad-spectrum resistance to diseases and insects. An example for the combination of two resistance QTLs is the pyramiding of a major stripe rust resistance gene and two QTLs in the same genotype. In order to pyramid disease or pest resistance genes that have similar phenotypic effects, and for which the matching races are often not available, MAS might even be the only practical method—especially where one gene mask the presence of other genes. Phenotypic selection in case of disease resistance is difficult because the gene being stacked control the same trait. Thus, MAS offers promising opportunities. At the same time, pyramiding has to be repeated after each crossing, because the pyramided resistance genes are segregating in the progeny.

Advantages of MAS over Conventional Breeding Protocols

- ☆ Distinguishing between heterozygote and homozygote genotypes: Visual selection is seldom effective in distinguishing between homozygous and heterozygous genotypes. Ability to make this distinction is required at some steps in some breeding programs.

- ☆ Early generation discrimination: MAS allows breeders to eliminate undesirable genotypes early in the breeding program by screening plants in the seedling stage.
- ☆ Convenient screening.
- ☆ Reduced space for screening
- ☆ Reduced breeding time. Early generation testing and ease of screening during selection may hasten the duration of the breeding program.

Limitations of MAS

1. Resources (equipment) not available
2. Markers may not be cost-effective
3. Most DNA markers have been developed for major genes, not QTLs. QTLs are much harder to characterize. (An exception is *Sub1*)
4. Accuracy of QTL mapping studies : QTL effects may depend on genetic background or be influenced by environmental conditions
5. Lack of marker polymorphism in breeding material
6. Poor integration of molecular genetics and conventional breeding

Recent Achievements in Marker Assisted Plant Breeding

- ☆ **Rice**: Improved Pusa Basmati 1(xa13 and Xa21), Improved Sambha Mahsuri (xa5, xa13 and Xa21), Improved Lalat and Tapaswini (Xa4, xa5, xa13 and Xa21), Swarna Sub1, DRR Dhan 42.
- ☆ **Wheat:** PBW 761 (Yr15), PBW 752(Yr10), PBW 723(Yr17, Yr40,Lr37 and Lr57).
- ☆ **Maize:** Vivek QPM9(opaque2), Pusa HQPM-7 Improved (crtRB1 and lcyE).
- ☆ **Pearl millet:** HHB 67 Improved and HHB 67 Improved 2 (DM resistance)
- ☆ **Chickpea:** Super Annigeri-1, Pusa chickpea 20211 (Fusarium wilt resistance)
- ☆ **Groundnut:** Girnar 4 and 5 (Oleic acid)
- ☆ **Soyabean:** NRC 142, 132 (less beany flavour), NRC 138 (Early maturity)

Conclusions and Future Prospects of MAS

Despite the relatively small adoption of markers in crop breeding to date, it is hoped that there will be a greater level of adoption within the next decade and beyond. Factors that should lead to a greater adoption of MAS in crop breeding include:

- ☆ Establishment of facilities for marker genotyping and staff training within many crop breeding institutes in different countries

- ☆ Currently available (and constantly increasing) data on genes/QTLs controlling traits and the identification of tightly linked markers
- ☆ Development of effective strategies for using markers in breeding units and curation of public databases for QTL/marker data
- ☆ Available resource for generating new markers from DNA sequence data arising from rice genome sequencing and research in functional genomics.
- ☆ Greater integration between molecular geneticists and plant breeders (and scientists from other disciplines)

It is also critical that future endeavours in MAS are based upon lessons that have been learnt from past successes and (especially) failures in using MAS. Further optimization of marker genotyping methods in terms of cost-effectiveness and a greater level of integration between molecular and conventional breeding (especially in designing efficient and cost-effective strategies) represent the main challenges for the greater adoption and impact of MAS crop breeding in the near future.

REFERENCES

Babu, N.N., Vinod, K.K., Krishnan, S.G., Bhowmick, P.K., Vanaja, T., Krishnamurthy, S.L., Nagarajan, M., Singh, N.K., Prabhu, K.V. and Singh, A.K., 2014. Marker based haplotype diversity of Saito/QTL in relation to seedling stage salinity tolerance in selected genotypes of rice. *Indian Journal of Genetics and Plant Breeding*, 74(01): 16-25.

Beckman, J. S. and Soller, M. 1983. Restriction fragment length polymorphisms in genetic improvement: methodologies, mapping and costs. *Theoretical and Applied Genetics* 67: 35-43.

Collard, B. C. Y., Jahufer, M. Z. Z., Brouwer, J. B. and Pang, E. C. K. 2005 An introduction to markers, quantitative trait loci (QTL) mapping and marker-assisted selection for crop improvement: the basic concepts. *Euphytica*, 142: 169–196.

Collard, B.C. and Mackill, D.J., 2008. Marker-assisted selection: an approach for precision plant breeding in the twenty-first century. *Philosophical Transactions of the Royal Society B: Biological Sciences*, 363(1491): 557-572.

Ganal, M. W., Altman, T., Röder, M. S. 2009. SNP identification in crop plants. *Current Opinion in Plant Biology*, 12: 211–217.

Gupta, P. K., Rustgi, S. and Mir, R. R. 2008. Array-based high-throughput DNA markers for crop improvement. *Heredity*, 101: 5-18.

Lateef, D.D. (2015) DNA Marker Technologies in Plants and Applications for Crop Improvements. *Journal of Biosciences and Medicines*, 3, 7-18.

Malik, W., Ashraf, J., Iqbal, M.Z., Ali Khan, A., Qayyum, A., Ali Abid, M., Noor, E., Qadir Ahmad, M. and Hasan Abbasi, G., 2014. Molecular markers and cotton genetic improvement: current status and future prospects. *The Scientific World Journal*, 2014.

Semegan K., Bjornstad A. and Ndjiondjop M. N. 2006. An overview of molecular marker methods for plants. *African Journal of Biotechnology*, 5 (25): 2540-2568 29.

Singh, B.D. and Singh, A.K., 2015. Marker-assisted plant breeding: principles and practices.

Vinod, K.K., Krishnan, S.G., Babu, N.N., Nagarajan, M. and Singh, A.K., 2013. Improving salt tolerance in rice: looking beyond the conventional. Salt stress in plants: signalling, omics and adaptations, pp.219-260.

Yadava,D.K., Hossain,F., Choudhury,P.R., Kumar, D., Singh, A.K., Sharma,T.R. and Mohapatra, T. 2022. Crop cultivars developed through molecular breeding (Second Edition). Indian Council of Agricultural Research, New Delhi.

Chapter 14

The Genetically Modified Organism from the Breeder's Perspective

Prachi Mahla[1]*, *Garima Vaishnav*[2] *and Anju Nehra*[3]

[1]*Ph.D. Scholar, Department of Genetics and Plant Breeding, MPUAT, Udaipur, Rajasthan*
[2]*Ph.D. Scholar, Department of Genetics and Plant Breeding, RARI, Jaipur, Rajasthan*
[3]*Ph.D. Scholar, Department of Genetics and Plant Breeding, SKNAU, Jobner, Rajasthan*
**e-mail: prachimahla001@gmail.com*

ABSTRACT

The world population is expected to reach nine billion people by 2050. This presents agriculture with a major challenge to feed the growing population in a sustainable manner. While the demand for food, including both grains and animal protein, is rising, agricultural yields for the majority of principal crops are now stagnant. It needs a persistent effort for producing a steady supply of superior cultivars for all the staple crops to meet the challenge of increasing yields. Conventional breeding cannot keep up with the growing demands; biotechnology and the development of genetically modified (GM) crops are filling the void. The safety of GM crops for the environment and human consumption is still under debate, though. The potential for GM technology to responsibly and sustainably fulfil future global demands for food, fibre, and feed is quite promising. To achieve the desired yields, nutritional quality, and sustainability, we must use all available tools, including conventional and organic food production systems. This chapter discusses the necessity for GM crops, how they are produced, their effects, and their safety.

Keywords: *GM crops, Agrobacterium, Sustainability, Safety concerns.*

Introduction

1. What are GMOs?

An organism (animal, plant, bacteria, or virus) whose genetic makeup has been altered for a specific purpose is referred to as a genetically modified organism (GMO). The organism does not naturally exist in this altered form. For instance, a plant can be altered by transferring a gene from another living organism, which modified to carry an additional gene found in another living thing (such as a bacteria) to protect itself against insect pests. DNA recombinant technology allows the transfer of genes from one organism to another, usually unrelated one. Genetically modified crops (GM crops) are agricultural plants whose DNA has been altered through the application of genetic engineering techniques. Plant genomes can be modified physically or by transferring sequences contained in T-DNA binary vectors using *Agrobacterium*. Tobacco was the first crop plant to be genetically modified in 1983. Over the past 25 years, GM crop production has increased more than 100-fold. Farmers currently cultivate over 190 million hectares of transgenic crops, which is roughly similar to Mexico's whole surface area. The four principal genetically modified agricultural crops are soybean (50 per cent), maize (30 per cent), cotton (13 per cent) and canola (5 per cent) (Figure 14.1).

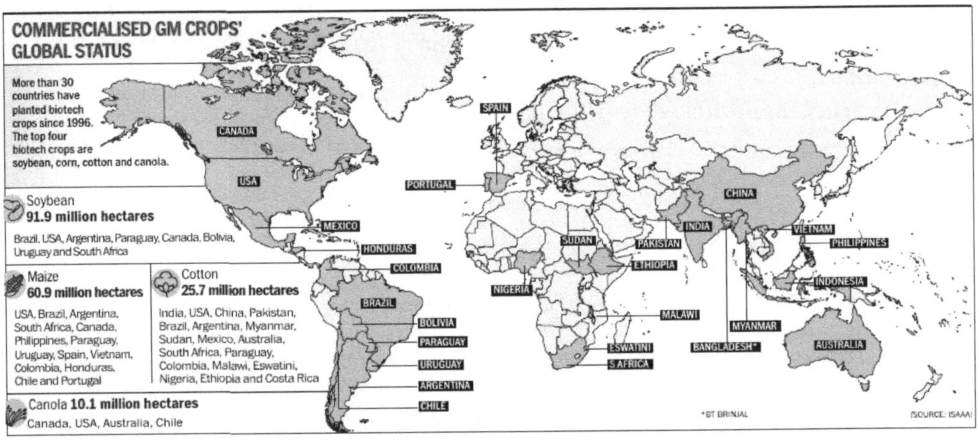

Figure 14.1: Global Status of GM Crops (Adapted from ISAAA).

2. How are GM Crops Produced?

A GMO is generally created by transferring genes into a crop plant over the course of two steps, first, successful gene transfer into plant cell, known as transformation and then regeneration of a transgenic plant from the transformed cell, typically in tissue culture. Gene transfer is often accomplished either through direct transfer of DNA or indirectly using a vector such as *Agrobacterium tumefaciens*.

Direct Gene Transfer

Several methods have been used to deliver DNA directly into plant cells, including electroporation into protoplasts, microinjection, chloroplast transformation, silicon-carbide slivers (coated in DNA), mesophorus silica nanoparticles and microparticle bombardment [1]. So far particle bombardment is the most commonly used method for direct DNA transfer. Sanford invented microparticle bombardment, also known as biolistics or the "gene gun," in the late 1980s, by firing pressurised helium-coated gold or tungsten microparticles (diameter between.5 and 1.0 m) coated with the engineered gene of interest as naked DNA into plant tissue at high velocities [2]. The pressure used to transfer the microparticle depends on the target tissues and can be as high as 2200psi. High velocity of the particles leads to deeper penetration into the tissues. Microparticle bombardment has been employed successfully to generate transgenic plants in a variety of crops, including all cereals, several tuber crops, and trees. It has the advantage of being able to transmit big DNA fragments and has even been utilised to transfer complete chromosomes and several independently created genes at the same time [3].

Indirect Gene Transfer

Although microparticle bombardment has been effective, Agrobacterium tumefaciens has far outpaced it in terms of commercial application. The soil bacterium A. tumefaciens infects plants, typically at the point where the stem and roots converge (referred to as the crown in gardening terminology), and changes the genome of the plant by introducing genes that promote cell multiplication. The bacteria live inside a mass of cells called a gall or tumour that is created by

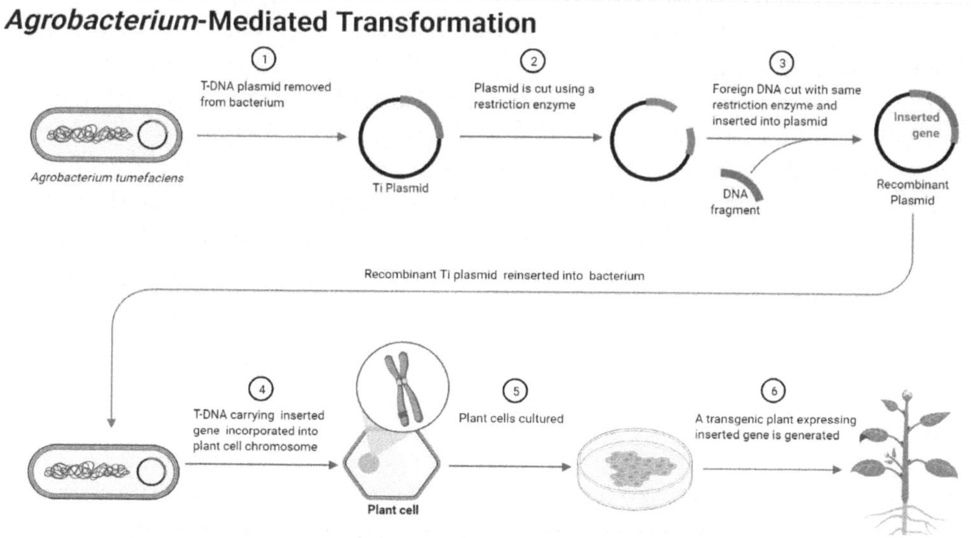

Figure 14.2: Indirect Gene Transfer through Agrobacterium.

the cell expansion. In addition to instructing the plant to develop a gall (crown-gall disease), A. tumefaciens also gives the cells the genetic instructions to produce opines, modified amino acids that the bacteria use as a source of food.

A tumour inducing, Ti plasmid contains a section of DNA *i.e.*, T-DNA which helps in gene transfer. T-DNA is flanked by two 25-base pair direct repeats on both right and left border and act as insertion site for T-DNA transfer into the plant genome. T-DNA contains genes that encode enzymes which synthesize auxin and cytokinin that cause tumour formation. Besides T-DNA, Ti plasmid contains the virulence genes that control the ability of the bacteria to infect the plant cell and transfer of T-DNA into the target plant cell [1].

GMO Applications

Genetically modified organisms have been used for a variety of medical, agricultural, environmental (bioremediation), and more recently, industrial purposes. These organisms include plants, animals, cellular bacteria, and viruses. The majority of GMOs that are intentionally released into the environment are GM plants used in agriculture. GM crops are grown in variable proportions in several nations, with the United States, Canada, Brazil, Argentina, and India being the leading producer. Maize, soybean, cotton, and rapeseed are the most common GM plants used in agricultural and industrial activities. Herbicide tolerance (specifically glyphosate and glyphosinate tolerance) and insecticide tolerance are the main characteristics introduced (mainly Bt or *Bacillus thuringiensis*) in GM crops. Additionally, recently, double-stacked and triple-stacked traits have been introduced.

1. Herbicide Tolerance

More than 90 per cent of corn, soybean and cotton grown in the United States are GM and herbicide-tolerant. As more nations adopt GM technologies, the area for the cultivation of herbicide-tolerant GM crops will increase. 98 per cent of the canola crop in Canada is genetically modified. The alleged problems with herbicide-tolerant plants include the emergence of "superweeds" (herbicide-resistant weeds), transgene flow (gene flow) to nearby non-GMO crops or wild relatives, and environmental/ecological issues with biodiversity and chemical use [4]. The GM herbicide tolerant crops have also been developed to enable the use of less toxic and more environmentallyfriendly chemicals. Glyphosate and glufosinate (another GM targeted herbicide resistance), for example are both Class III herbicides (EPA), which are only slightly toxic and have low persistence in the soil and environment, averaging approximately 40 days. These herbicides have replaced herbicides either more toxic or that is known to contaminate and persist in groundwater [5].

2. Insect Resistance

Insect-resistant GM crops have also been widely adopted in the United

States; over 90 per cent of maize, soybeans, and cotton are GM for insect resistance, and insect-resistant GM crops, like herbicide-tolerant GM crops, are rapidly expanding in area globally [6]. For insect resistant GM crops, the transgene which allows the synthesis of CRY protein toxin from the bacteria *Bacillus thurengensis* is used. This toxin affects the gut of the insect and prevents digestion. CRY toxins are selective to their target insects, are harmless to vertebrates, including humans, and have no effect on the plant. They are also biodegradable, which means they will not linger in the environment [7]. Because of this, GM technology can be used to address insect pests and the damages and yield losses they cause. There is no harmful effect of Bt crops on environment and human health as the toxin in the transgenic crops only affect the targeted insects.

3. Other Applications

Genetically modified seeds have the potential to boost agricultural production or, at the very least, preserve the maximum yield of vegetation. Increased yields can be achieved by using GMO plants with insect and herbicide tolerance, which considerably simplify crop management and cut crop losses. Commercialized genetically modified plants often have lower pesticide and tillage requirements, which have positive environmental effects. Reducing the use of pesticides can conserve species that aren't the intended targets while also increasing the conservation of beneficial insects. Reduced tillage reduces soil erosion and pollution while also having indirect environmental advantages like less pesticide and fertiliser runoff contaminating the water supply [8].

GMOs help farmers to spend less time on assets, which allows them to generate more income. Additionally, it is economical because there is no need for pesticides because these organisms are created to be pest resistant. The purpose of genetically modified organisms is to produce more vitamins, minerals, and other nutrients [9]. For instance, scientists were able to create a modified form of African corn that contains 169 times as much beta-carotene as regular plants, 6 times as much vitamin C, and 2 times as much folate as standard crops. This is especially helpful in areas where there are dietary deficiencies among the population. Several GMO crops have been modified to reduce their susceptibility to insects and pests. In addition to posing fewer health risks to those who consume the vegetation, using fewer pesticides may also have a positive impact on the environment.

Benefits of GM Crops

- ☆ Environmental benefits: Farmers will be able to use fewer pesticides on insect-resistant GM plants. This ensures that less potentially harmful substances are absorbed into the soil.
- ☆ Since insect-resistant plants require less insecticide to grow, farmers will produce healthier food.

Crop	Trait	Varieties	Year of release	Releasing organizations/ country
Insects/Pest Resistance				
Cotton (*Gossypium hirsutum* L.)	Bt-(Cotton Bollworm, tobacco bud worm and pink bollworm)	Bollgard	1995	Monsanto
	Bt gene	NA	NA	Calgene
Potato (*Solanum tuberosum* L.)	*Bt* gene (Colorado potato beetle)	New Leaf	1995	Monsanto
Maize (*Zea mays* L.)	*Bt* gene (Corn borer)	Maximizer	1995	Northrup King (Sandoz)
	Bt gene (Corn borer)	Yieldgard	1996	Monsanto
	Bt gene (Corn borer)	NA	NA	Northrup King
				Novartis
				Mycogen
	Corn Root Worm	MIR604	2003	Monsanto
Squash (*Curcubita pepo* L.)	Virus resistance	Freedom	NA	Poineer Hi Bred
				Asgrow Seeds
Tobacco (*Nicotiana tabacum* L.)	Virus resistance	NA	NA	China
Tomato (*Lycopersicon esculentum* (L) Karsten)	Virus resistance	NA	NA	China
Capsicum (*Capsicum annuum* L.)	Virus resistance	NA	NA	China
Cantaloupe (*Cucumis melo* L. subsp. *melo* var. cantalupensis Naudin)	Virus resistance	NA	NA	Asgrow Seeds
Herbicide Resistance				
Maize (*Zea mays* L.)	Glufosinate	Libertylink	1996	AgrEvo
	Sethoxydim	SR	NA	BASF
	Glufosinate	NA	1996	Dekalb Genetics
	Glyphosate	Roundup Ready	NA	Monsanto
	Glufosinate	NA	NA	Plant Genetics Systems
Conola (*Brassica napus* var. *napus* L.)	Glufosinate	Inovator	NA	AgrEvo
	Glyphosate	Roundup Ready	NA.	Monsanto
	Bromoxynil	NA	NA	Rhone-poulenc
Cotton (*Gossypium hirsutum* L.)	Bromoxynil	BXN	1995	Calgene
	Sulfonyl urea	NA	1996	Du Pont
	Glyphosate	Roundup Ready	1996	Monsanto
Soybean (*Glycine max* (L.) Merr.)	Glyphosate	Roundup Ready	1995	Monsanto
Flax (*Linum usitatissimum L.*)	Sulfonyl urea	Triffid	NA.	Canada
Quality				
Tomato (*Lycopersicon esculentum* (L) Karsten)	Vine ripened shelf life (Antisense RNA Technology)	FlavrSavr	1994	Calgene
	Vine ripened shelf life (Antisense RNA Technology)	Endless & Summer	1995	DNA Plant Technology
	Delayed Softening	NA	1995	Zeneca
	Thick skin paste consistency			
	Delayed ripening	NA	1995	Monsanto
Potato (*Solanum tuberosum* L.)	Modified starch	NA	NA	AVEBE
Conola (*Brassica napus* var. *napus* L.)	Oli characteristics (High lauric acid)	Laurical	1995	Calgene
	Male sterility hybrid system			AgroEvo

Source: Kumar et al.;[121] NA, Information not available.

Figure 14.3: Genetically Engineered Varieties Developed in different Crops (Adapted from Kumar *et al.*, 2010)

- ☆ GM crops are more resistant to diseases, insects and herbicides, resulting in a more constant production.
- ☆ An Insect resistant crop reduces the requirement of insecticides to be sprayed on entire crop, which affects whole insect population. While, insect resistant GM crop only affects the targeted pest.

- Reduced demand of pesticides and insecticides lowers the total cost of cultivation.
- No-tillage farming, which has advantages for both the farmer and the environment, may be achieved with GM crops.

Limitations of GM Crops

- Genetically modified crops may cause food allergies due to foreign genes in the food and may cause health related problems.
- Genetically engineered products may contaminate the non-genetically engineered products or their end products.
- When gene engineers splice a foreign gene into a plant or microbe, they frequently link it to another gene known as an antibiotic resistance marker gene (ARM). These ARM genes could unexpectedly recombine with disease-causing bacteria or microorganisms in the environment or in animal or human guts, resulting in antibiotic resistance.
- Due of the instability of transgenic traits, they may degrade and revert flower development, spread to native trees, or produce pollen that is toxic to bees and other pollinators as well as potentially harmful to humans.
- The purposeful or unintentional introduction of modified genes into native populations may disturb natural ecosystems and biodiversity.
- "Superweeds" and "Superpests": These are crops that have been genetically modified to be herbicide resistant or to create their own pesticide. Common plant pests can evolve into superpests that are fully immune to pesticides like BT-sprays and other environmentally sustainable biopesticides if they are constantly under pressure from genetically altered crops.
- New viruses and bacteria can be created by genetically engineering plants to resist viruses, which can drive viruses to mutate into new, more aggressive forms.

Safety Concern Related to GM Crops

The production of GM crops is contentious, owing to concerns about the long-term impact on human health and the environment. GM foods must pass strict safety tests. Only GM foods are exposed to such stringent testing. There are few major concerns related to safety of GM crops:

- Outcrossing refers to the transfer of genes from GM crops to traditional crops or wild relatives. This can be avoided by clearly dividing GM and conventional crop fields.
- Although the risk is almost negligible and not limited to GM foods, genes associated with GMOs have the potential to be transferred to

the body or the bacteria that naturally live in the gut. This has the potential to produce superbugs.

☆ The transfer of genes from foreign organisms has the potential to create new allergies. Allergens are substances that trigger allergic reactions. There have been no reports of allergic reactions to any of the GM foods currently on the market. For example, during the creation of a GM soybean crop, researchers got concerned when it was discovered that the soybeans triggered an allergic reaction. This soybean crop was genetically modified to include a gene from a Brazil nut in order to increase the protein content of soybeans. The Brazil nut is a highly allergenic food, and it was later determined that the allergic reaction was induced by the gene transfer.

Current Status of GM Crops in India

The Government of India's GEAC (Genetic Engineering Approval Committee) approved the commercial cultivation of genetically engineered cotton (commonly known as Bt cotton) for pest resistance in 2002 and it has resulted in a paradigm shift in Indian GM crop research, deregulation and even the cotton sector in India. Since then, cultivation and production of Bt cotton have increased tremendously, and India has emerged to become the world's second largest cotton producer and exporter. The GEAC approved Bt brinjal for commercialization in October 2009, but due to concerns raised by some farmers, anti-GM activists, and scientists, the government of India officially declared a temporary ban on 9 February 2010, and then Environment and Forest Minister Mr. Jairam Ramesh stated that there is no overriding urgency to introduce Bt brinjal in India and also reiterated that the government had only imposed a moratorium. Despite all of these achievements, India remains primarily an agricultural economy, with about 70 per cent of the working population relying on agriculture for employment and subsistence, and farmer conditions are deplorable.

The story of Bt cotton producers is not as cheery as several scientists have reported, and if there is no clear evidence that GM crops can actually increase yields, then there is no urgent need to plough ahead for commercialization of GM crops in India and wish to develop new policy for ensuring food security in the future without compromising the security of human and livestock health and jeopardising the vast majority of farmers. The development process, potential benefits, and deregulation of Bt Brinjal in India are among the most contentious issues. On July 18, 2014, the GEAC (Genetic Engineering Appraisal Committee, established on July 22, 2010) approved limited experimental field trials of GM rice, brinjal, mustard, cotton, and chickpea for the sole purpose of gathering biosafety data. The GEAC has recommended commercial cultivation of GM mustard for clearance to the Environment Minister (Indian Express 2017), but the final decision is still awaited.

Conclusion

Genetic modification refers to the introduction of a gene that has a positive outcome. Public discussion has been sparked by the marketing of GM foods and the discharge of GMOs into the environment in various industry sectors. According to the studies mentioned here, it would be foolish to discount GMO vegetation as a means of achieving goals for sustainable global growth at a time when the sector's population is rising. GMOs benefit humanity when used to achieve objectives including increasing the accessibility and quality of food and healthcare, as well as helping to maintain a cleaner environment. Genetically modified crops are more resistant to abiotic challenges, rely less on chemical insecticides, use minerals more effectively by agricultural plants, experience less post-harvest loss, and have longer dietary costs of food.

REFERENCES

1. Barampuram S and Zhang ZJ. Recent advances in Plant transformation. In: Plant Chromosome Engineering: Methods and Protocols. Ed: J. A. Birchler, Methods in Molecular Biology, Humana Press. 2011; 701: 1-35.
2. Sanford JC. Biolistic plant transformation. Physiol. Plant. 1990; 79: 206-209.
3. Schmidt MA, Lafayette PR, Artelt BA and Parrott WA. A comparison of strategies for transformation with multiplegenes via microprojectile-mediated bombardment. In Vitro Cell Dev. Biol. Plant. 2008; 44: 162-168.
4. Busi R, Vila-Aiub M M, Beckie H J, Gaines TA, Goggin DE, Kaundun S S, *et al.*, Herbicide-resistant weeds: from research and knowledge to future needs. Evolutionary Applications. 2013; 6: 1218-1221.
5. Roland P. Plant genetics, sustainable agriculture and global food security. Genetics. 2011; 188: 11-20.
6. James Clive. Global status of commercialized biotech/GM Crops: 2013. ISAAA Brief No. 46. ISAAA: Ithaca, NY.
7. Bravo A, Gill SS and Soberón M. Mode of action of Bacillus thuringiensis Cry and Cyt toxins and their potential for insect control. Toxicon. 2007; 49: 423-435
8. Chandler S and Dunwell JM. Gene flow, risk assessment and the environmental release of transgenic plants. Crit Rev Plant Sci. 2008; 27: 25-49
9. Kausch AP, Hague J, Deresienski A, Tilelli M and Nelson K. Male sterility and hybrid plant systems for gene confinement. In: Plant Gene Containment.

Eds: M.J. Oliver and Y. Li. Wiley-Blackwell Publishing Ltd. Hoboken, NJ. 2012; 85-100

10. Basu SK, Dutta M, Goyal A, Bhowmik PK, Kumar J, Nandy S, *et al.*, Is genetically modified crop the answer for the next green revolution? GM crops. 2010; 1(2): 68-79.

Chapter 15

Resistance Breeding and its Current Updates

M.O. Ankitha, S. Anand and S. Bhaskar Reddy*

PhD Scholar, Department of Plant Breeding and Genetics, College of Agriculture, Vellayani, Kerala Agricultural University

****e-mail: ankithamo42@gmail.com***

Introduction

To fulfil the predicted food demand of the world's 10 billion inhabitants in 2050, global food production must rise by at least 60 per cent. (Food and Agriculture Organization, 2019[1]). Crop production must be significantly increased in order to achieve this goal. Plant diseases and pests are thought to be responsible for a global crop loss of 17–30 per cent, depending on the crop (Savary *et al.*, 2019[2]). Crop losses can be managed by resistance breeding in an economical, environmentally friendly, and labor-saving way. Inherited resistance is a useful quality since it is simple for the grower to use and decreases the need for additional management measures, such as the application of chemicals. Resistance breeding includes changing a plant's genetic makeup to create resistant cultivars, which is a less expensive, more practical, and environmentally beneficial option.

Breeding varieties/hybrids resistant to biotic stresses *viz.*, pests, diseases and nematodes and abiotic stresses *viz.*, drought, salinity and adverse climatic conditions like frost, chilling temperature are the primary objectives in any breeding programme. However, it is challenging to breed varieties with resistance that is effective, stable and broad-spectrum. Recent developments in genetic and genomic technology have helped to improve our comprehension of the intricacy of host-pathogen interactions and have helped to identify some of the genes and mechanisms underlying resistance. This new information is assisting crop improvement through better-informed breeding techniques that make use of various forms of resistance at various scales, from the genome of a single plant to the plant varieties used throughout a region.

Biotic Stress

Biotic stress is stress that occurs as a result of damage done to plant by other living organisms, such as bacteria, viruses fungi, parasites, beneficial and harmful insects, weeds and cultivated or native plants. Plant resistance to diseases can be classified in to vertical resistance and horizontal resistance. Vertical resistance is the resistance of a host to the particular race of pathogen; also called as major gene resistance, oligogenic resistance, qualitative resistance. Horizontal resistance refers to the resistance of a host to all the prevalent races of a pathogen; also called as minor gene resistance, polygenic resistance, nonspecific resistance and general resistance (Singh 1983[3])

Mechanism of Disease Resistance

1. Disease escape: The ability of susceptible host plants to avoid attack of disease due to environmental conditions factors, early varieties, charge in the date of planting, change in the site of planting; balanced application of NPK *etc.* is called disease escape.
2. Disease endurance or tolerance: The ability of the plants to tolerate the invasion of the pathogen without showing much damage. This endurance is brought about by the influence of external characters. Generally, tolerance is difficult to measure since it is confounded with partial resistance and disease escape.
3. Immune reaction: When the host does not show the symptoms of disease it is known as immune reaction. Immunity may result from prevention of the pathogen to reach the appropriate parts of the host *e.g.* exclusion of spores of ovary infecting fungi by closed flowering habit of wheat and barley.
4. Hypersensitivity: Immediately after the infection several host cells surrounding the point of infection are so sensitive that they will die. This leads to the death of the pathogen because the rust mycelium cannot grow through the dead cells. This super sensitivity (hypersensitivity) behaves as a resistant response for all practical purposes. Phytoalexins are specific polyphenolic or terpenoid chemicals and are produced by the host in response to the infection by a pathogen.

Sources of Disease Resistance

- ☆ A known variety: Disease reactions of most of the cultivated varieties are documented and a breeder may find the resistance he needs in a cultivated variety.
- ☆ Germplasm collection: When resistance to a new disease or a new pathotype of a disease is not known in a cultivated variety germplasm collection should be screened.
- ☆ Related species: Often the resistance to a disease may be found in related species and transferred through interspecific hybridization. *e.g.* Resistance to stem, leaf and stripe rusts of wheat.
- ☆ Mutation: Resistance to diseases may be obtained through mutation arising spontaneously or induced through mutagenic treatments. Eg, Resistance to Victoria blight in oats was induced by irradiation with x-rays or thermal neutrons/also produced spontaneously, Resistance to stripe rust in wheat, Resistance to brown rust in oats, Resistance to mildew in barley.

Mechanism of Insect Resistance

- ☆ Antixenosis (Non-preference): Antixenosis is mechanism employed by the plant to deter or reduce colonization by insects. Plants that exhibit antixenotic resistance should have a reduced initial number of colonies early in the season.
- ☆ Antibiosis: Antibiosis operates after the insect have colonized and have started utilizing the plant.
- ☆ Tolerance : Tolerance of the plant does not affect the rate of population increase of the target pest but does raise the threshold level.

Abiotic Stresses

The potentially harmful impacts of salinity, drought, flooding, metal toxicity, nutrient insufficiency, high temperature, and low temperature are examples of abiotic stresses. Abiotic stresses can also include infrequent and extremely localised conditions like Shade, UV exposure, Photoinhibition, Air pollution, Wind, Hail, and Gaseous shortage. Abiotic stress can occur to plants due to a lack of a necessary resource or the presence of a poisonous or antagonistic agent in high concentrations. In some circumstances, such as the availability of water, stress can be caused by both too much (flooding) or too little (drought).

Drought Resistance

The ability of crop plants to grow, develop and reproduce normally under moisture deficit conditions is referred to as drought resistance.

Mechanisms of Drought Resistance

- ☆ Drought escape: Plant avoid the injury of stress by regulating its life cycle to avoid meeting with stress. This is not the kind of resistance – some short-lived, desert ephemeral plants germinate, grow and flower very quickly following seasonal rains. They thus complete their life cycle during a period of adequate moisture and form dormant seeds before the onset of dry season.
- ☆ Drought avoidance: Drought avoidance refers to ability of the plant to maintain a favourable internal water balance under moisture stress. In other words, plants which avoid drought retain high water contents in their tissues. Drought avoidance can permit a longer growth period in the crop through reduced water use or increased water uptake
- ☆ Drought tolerance: The ability of crop plants to withstand low tissue water content is referred to as drought tolerance. Drought tolerance is more desirable because the crop can produce more yield at lower water potential.
- ☆ Drought resistance: Drought resistance is the sum of drought avoidance and drought tolerance. It refers to the ability of crop plants to give higher yield under moisture stress conditions or survival of plants under water deficit conditions without injury.

Salt Tolerance

Ability of plants to prevent, reduce or overcome injurious effects of soluble salts present in their root zone

Mechanism of Salt Tolerance

- ☆ Salt tolerance: By accumulating salt, generally in their cells or glands and roots. Halophytes show tolerance by ion accumulation mechanism.
- ☆ Salt avoidance: By maintaining their cell salt concentration unchanged either by water absorption (*e.g.* Rice, chenopodiaceae) or by salt exclusion (*e.g.* tomato, Soya bean, citrus, wheat grass) Glycophytes (nonhalophytes) owe their resistance primarily to avoidance *e.g.* barley

Chilling Tolerance

- ☆ Chilling: When temp remain above freezing *i.e.* >0°C to < 10-15°C.
- ☆ Freezing: When temp remains below freezing *i.e.* <0°C.

Mechanism of Chilling Resistance

- ☆ Chilling resistance: Chilling sensitive plants are typically tropical plants. Temperate plants generally tolerate chilling injury.

- Chilling tolerance mechanisms involve: Membrane lipid un-saturation, Reduced sensitivity of photosynthesis, Increased chlorophyll accumulation, Improved germination, Improved fruit/seed set,

Mechanism of Freezing Resistance

- Freezing avoidance: The ability of plant tissues/or genes to avoid ice formation at sub zero temperature Super cooling is a mechanism of freezing avoidance which is controlled by Lack of ice nucleators, Small cell size, Little or no intercellular space, Low moisture contents Barriers against external nucleators
- Freezing tolerance: Ability of plants to survive the stress generated by extra cellular ice formation and to recover and re grow after thawing

Resistance Breeding

Breeding approaches for biotic and abiotic stress resistance can be broadly classified as two, *i.e.* conventional and molecular breeding methods. Conventional approach includes pureline, pedigree, backcross, recurrent selection and interspecific hybrids through which various resistant varieties have been developed. The continual sowing of these kinds, however, causes abrupt outbreaks of pests and illnesses after a while, which prompts advances in molecular biology. The discovery of trait-linked markers also paved the way for the development of disease-resistant varieties. Gene deployment and multiline varieties, which are heterogeneous populations made by combining cultivars to differ in their ability of disease resistance and slow the rate of epidemic development, are examples of molecular approaches. Gene pyramiding involves introducing two or more genes that provide resistance to various types of diseases and pests into a particular variety. Several connected markers that confer resistance against diseases like wheat stem rust and leaf rust are identified with the aid of makers in marker aided breeding employing foreground, recombinant, and background selection. With the advancement of more sophisticated technologies, transgenic procedures were created, which use RNA-mediated silencing to silence genes and manage a number of bacterial, viral, and fungal infections. In addition, through modifying different defence mechanisms, forward breeding and reverse breeding also contribute significantly to the control of diseases. Last but not least, tissue culture plays a crucial part in disease management. There are several different toxins and culture filtrates that have been created that offer resistance to many diseases. Sugarcane varieties like Ono, which are resistant to Fijian diseases, have been released through somaclonal variations.

1. Conventional Approach

It is a long-standing set of techniques that dates back to the dawn of mankind. Numerous cultivars have been evolved from the dawn of human civilization to the present day. The most well-known examples include the creation of hybrid rice in the 1970s and semi-dwarf, high yielding grain cultivars during the Green Revolution. Some of them are:

1. Introduction: plants introduced from their native place to another place for crop improvement. Eg; Early varieties of groundnut introduced from USA have been resistant to leaf spot (Tikka). Kalyanasona and Sonalika wheat varieties originated from segregating material introduced from CIMMYT, Mexico, were rust resistant. African bajra introductions have been used in developing downy mildew resistant cms lines.
2. Selection: The oldest and most fundamental method of breeding plants is selection. The simplest and least expensive way to create a resistant variety is to select biotic resistant plants from a commercial variety. Although this approach has proven effective in numerous situations in the past, it is only marginally useful at the current stage of crop improvement. For example, Kufri Red potato is a disease resistant selection from Darjeeling Red Round. Other such examples include resistance to curly-top in sugar beets, to mildew and leaf spot in alfalfa, to cabbage yellows in cabbage and to Periconia root rot in sorghum.
3. Pure line method : It is developed especially for the self pollinated crops. Pureline cultivars are desired in areas where uniformity of a product fetches a high premium because they are offspring of single, homozygous, self-pollinated species and can be easily developed in nature. They are also important in the breeding of inbred lines that are used to create hybrids in self or cross pollinated crops. Since selection is focused on progeny performance, this strategy allows for the greatest improvement over the original types while also being helpful in addressing low heritability features.
4. Pedigree method: Here detailed record of the origin of the selected individuals or lines are maintained. These records allow us to predict which parent superior F2 progeny will grow. It is helpful for highly heritable features such seed size, shape, disease, insect, height, and maturity. If controlled by important genes, it is frequently employed to create disease-resistant varieties (Miah *et al.*2013[4]). Laxmi, is a cotton variety resistant to leaf blight, which was developed using the pedigree approach. However, the varieties created as a result of it cannot be employed in settings where the genetic variation for the desired traits is not manifested, and it necessitates extensive testing and record keeping. Additionally, it is expensive in terms of the amount of labour and land needed. It is ineffective for characters controlled by polygenes.
5. Backcross breeding : Backcross varieties, which are helpful in small grain crop breeding, were created by crossing F1 hybrids with any of

the known parents. Using this technique, two plants are chosen and crossed. The progeny produced by regular backcrossing with one of the recurrent parents is nearly identical to the recurrent one. In this procedure, the desirable variety is known as a recurrent or receiver parent because it lacks certain desirable traits, whereas the unwanted variety, or wild variety, only possesses one or two desirable traits.

6. Interspecific hybridization: In order to pass on one or a few simply inherited traits, such as disease resistance, to a crop species, two species of the same genus are crossed to create an interspecific hybrid. Genetic variation of one or both parents increases through introgression. The potato variety Kufri jyoti, which is resistant to late blight, was created by interspecific hybridization. Through interspecific hybridization gene for resistance to cotton rust caused by *Puccinia cacabatahas* been transferred from *Gossypium anomalum* and *Gossypium arboreum* into *Gossypium hirusutum* (Anjum *et al.,* 2015[5]). A more diverse range of resistance can be achieved by using the highly effective technology of interspecies gene transfer.
7. Mutation breeding: Since mutation breeding is particularly effective at transferring disease resistance, it is utilised in addition to conventional breeding. Physical mutagens include (a) ionising radiations, which are further divided into non-particulate radiations (X-rays, gamma rays) and particulate radiations (alpha, beta, fast, and thermal neutrons). (b) Non-ionizing radiations (UV rays). Alkylating agents, base analogues (5-Bromouracil, 2-Aminopurine), acridine dyes (acriflavin, proflavin, acridine orange, ethidium bromide), deamination agents (nitrous acid), and other chemical mutagens are examples of chemical mutagens. Mustard or sulphur mustard, nitrogen mustard, ethyl methane sulphonate, methyl methane (hydroxylamine, sodium azide, DNA sequences). Fast neutrons, thermal neutrons, X rays, EMS, and sodium azides are some of the mutagens that are often utilised in plants, although gamma rays are presently the most frequently used. (Singh 2013[6])

2. Molecular Approach

This strategy was developed when molecular markers were discovered. There are several prospects for scientists to further advance contemporary cultivars thanks to the development of gene identification technologies using biotechnological tools. Breeders typically exploit major gene resistance to create new varieties, however over time, this method fails to provide long-term disease control, which results in epidemics (Gnanamanickam *et al.,* 1999[7]). Comparing morphological markers used in traditional plant breeding with molecular markers for trait selection reveals many advantages. This has led to the development of several sources with diverse levels of resistance to certain diseases. Recent developments in cellular and molecular biology have enabled techniques to create more resilient resistant cultivars and to solve the issue of the scarcity of donors for specific diseases (Khush, 2005[8]). Gene deployment, marker-assisted breeding, and gene pyramiding are a few techniques that are frequently employed to increase a cultivar's resistant durability.

(i) Gene Pyramiding

The breakdown of monogenic resistance, which involves the inclusion of two or more main genes in the host for specific resistance to the pathogen, gave rise to this novel idea. Horizontal resistance can be successfully used by gene pyramiding. One of the most well-known epidemic diseases that has struck and wrecked havoc on the economy, like the Irish famine brought on by *Phytopthora infestans*, causes a crop to completely collapse. It was therefore believed that pyramiding many disease resistance genes in a single cultivar might offer a long-lasting resistance, resulting in a decrease in the pathogen's rate of reproduction. Using regular backcrossing for the development of genes stacked variety takes a long period of time therefore discovery of markers makes the gene pyramiding somehow easier. Apart from that some of the bacterial blight resistant lines which have been developed in India are IR 24, Lalat and Tapaswani. Major problems associated with gene stacking is for finding resistance gene. However, if anyhow it is found then there is problem of linkage drag and many times it is also seen that there is adverse effect on yield of the crop.

(ii) Gene Deployment

Major genes are deliberately used in the production of resistant cultivars for distinct geographical regions. In the north of the US, it was originally applied to soybean breeding. The identification and characterization of many resistance genes has taken place. These genes provide a different way to introduce specialised resistance to crop improvement programmes, especially in conjunction with the creation of effective plant transformation systems. By blending cultivars with different levels of disease resistance and slower rates of epidemic emergence, heterogeneous populations are created (Borlaug, 1958[9]). In this regard, a strategy involves using various kinds in a predetermined manner and using each variety's unique resistance mechanism (temporal or spatial). Different location-specific cultivars are utilised depending on crop production techniques, seasonal preferences, and geographical preferences. As a result, various resistance sources have been used to create unique kinds. Variety with a variable maturation date, even within those utilised for a specific season, should include distinct sources of disease resistance. This will reduce the emergence of new virulent races and increase the resilience of resistance in current types. One method for enhancing the persistence of resistance in newly developing varieties is the specific gene deployment in various age groups. It assists in reducing selection pressure and consequently hinders the evolution of resistance in the pathogen population (Bonman, 1992[10]). Two challenges, however, are related to it. The first is that there is no guarantee of long-lasting resistance, and the second is that because of epistatic gene interaction, it may not always be possible to select for the presence of various genes or for particular gene combinations.

a) Spatial gene deployment: In this gene deployment is done over small geographical region like at the farm level, field level, regional level which includes variety mix ture, species mixture, multilines and multiblends (mixture of existing lines or varieties).Variety mixture includes homogenous spatial mixture of different genotypes of one plant species in a field (Adugna, 2004[11]). Resistance in a mixture can be obtained through decrease in inoculum potential, cross protection. According to Jorgensen, 1993 when a mixture of cultivar having different gene for disease resistance is used it provides a long term durable resistance. Oat and barley mixtures are used for controlling oat sterile dwarf virus.

b) Temporal gene deployment: It can be done in one of two ways: either by using a variety resistant to disease that has different genes for disease resistance, or by sequentially releasing a variety with those genes and then regularly replacing it with a new variety as new diseases emerge and have different genes for disease resistance.

(iii) Multiline Variety

It is the mixture of several similar purelines having different genes for disease resistance. For development of multiline varieties several purelines with different resistance genes should be developed through back cross programmes using one recurrent parent which is done by transferring disease resistance genes from several donor parents carrying different resistant genes to a single recurrent parent. Each donor parent is used in a separate backcross programme so that each line has different resistant gene. According to Borlaug, 1958 a total of eight to ten lines are used which require eight to ten backcrossing for development of multiline variety which helps in exploitation of horizontal resistance. The durability of resistance depends upon the rate of disease development, number of lines components in a mixture and the extent of planted area. According to Marshal, 1977[12] two approaches have been proposed for multiline varieties *i.e.* clean crop approach and dirty crop approach. According to clean crop approach, each donor has different genes for resistance to the given disease so that each pureline obtained from backcross progenies has a different resistance genes. If any component line becomes susceptible to the pathogen it is immediately replaced by a new resistant line. According to dirty crop approach, a multiline variety consists of several lines none of which is resistant to all the known races of the pathogen. Such a multiline variety is regularly attacked by pathogens but the spread of disease is expected to be much slower than in a pureline variety. Major shortcoming associated with multiline variety isfrom farmer's point of view as it is difficult for them to change the seeds on every few years depending upon the change in the races of pathogen and by the time a multiline is developed its recurrent parent is immediately replaced by the other superior varieties.

(iii) Marker Assisted Breeding

Different kinds of genetic markers have been produced in crop plants during the course of the 1980s and 1990s thanks to advancements in molecular biotechnology. Marker-assisted breeding was made possible by the rapid advancement of molecular markers (especially DNA markers) and the ongoing development of molecular assays (MAB). By performing selection not just on the characteristic of interest but also on associated markers of that particular trait, molecular markers have a larger potential to increase the efficacy of conventional breeding. There are other sorts of markers, but the SSR is the most popular because it is both reliable and co-dominant in nature (Kumar *et al.*, 2015[13]). MAB has a number of benefits over traditional breeding practises. (1) It allows selection for all kinds of traits at seedling stage and reduces the time required before the phenotype of an individual plant is known. (2) There is no effect of environment, thus allowing theselection to be performed under any environmental conditions. (3) Codominant markers allow effective selection of recessive alleles of desired traits in the heterozygous status. (4) For the traits controlled by multiple genes/QTLs, individual genes/QTLs in the same individuals can be identified and selected simultaneously and thus suitable for gene pyramiding. (5) There is also higher effectiveness and efficiency in terms of time and resources.

Through MAB genes for bacterial blight resistance like Xa 4, xa 5, xa 13 and Xa 13 has been incorporated into multiple background with non-additive effect combination of Xa 21 + xa 13 + xa 5, Xa 21 + xa 5, Xa 21 + xa 13 possess high level of resistance (Mishra *et al.*, 2016[14]).

(v) Transgenic Strategies

It is a sophisticated form of genetic engineering in which foreign genes are inserted into the plant species' genomes in order to achieve desired results. This technique allows for the simultaneous insertion of many genes into the genome to produce broad-spectrum resistance lines while allowing transgenic plants to acquire a single desirable characteristic.

(vi) Tissue Culture Approach

Tissue culture is crucial in the selection of explants, which are chosen either through callus culture or toxins. These toxins include phytotoxic metabolites produced by pathogens, which can include bacteria or fungi that are pathogenic.

Numerous experts have noted that these phytotoxins are regarded as an effective tool for selecting and breeding plants with disease resistance. The two types of these toxins are host-specific and non-specific toxins. The development of plant diseases is significantly influenced by two of these metabolites, fusaric acid and deoxynivalenol. In the case of culture filterate, disease-resistant plantlets are obtained by selecting survivor explants from altered culture filterate and regenerating them.

Recent Updates in Resistance Breeding

Multiple Disease Resistance by Speed Breeding

Plant breeders are experimenting with new approaches to improve crop production quality to respond faster to changing climates and emerging diseases. Lee T. Hickey *et al.*[15] obtained the two-row barley cultivar, Scarlett, with novel approaches for rapid trait introgression in a study. They developed 87 $BC_1F_{3:4}$ Scarlett introgression lines (ILs) in two years using four donor lines with multiple disease resistance and a redesigned backcross method that included phenotypic multi-trait screens as well as fast generation advanced technology 'speed breeding' (Bisht *et al.*, 2022[16]).

Genome-Wide Association Mapping for Biotic Stress Resistance

The genome-wide association study (GWAS) approach allows researchers to examine the enormous allelic diversity in natural crop germplasm. The millions of crossing events in germplasm during evolution increases the mapping resolution of GWAS. While GWAS is used widely for many plant species (Tanin *et al.*, 2022[17]), relatively few studies on the traits contributing to resistance to different biotic stresses have been reported in soybean

Genome Editing for Developing Resistant Cultivars

With the advent of large sequencing data, mapped QTLs, identified candidate genes and trait-associated SNPs and InDels, the architecture of plant genomes can be reshaped using genome editing approaches to suit the needs of the growing population. Among others genome editing tools, CRISPR/Cas9 has emerged as the most successful and widely applicable genome editing tool owing to its specificity and minimum off-targets. Genes involved in biotic stress responses can be edited using CRISPR/Cas9 to impart resistance and generate multiple sources of defense response to different pests and pathogens

Transcriptome Analysis

Transcriptome of protein-coding, non-coding, and regulatory genes is essential for comprehending the complex molecular mechanisms and gene regulatory networks governing resistance against different biotic stresses. Protein-coding genes include functional genes that are involved in downstream signaling related to defense-related pathways. In contrast, non-coding genes such as microRNAs (miRNAs), long non-coding RNAs (lncRNAs), and regulatory genes (TFs) regulate the expression of functional genes. Several methods, including northern blotting, microarrays, and RNA-sequencing (RNA-seq) analysis, have been used to conduct transcriptomic studies in soybean.

Consequently, microarray profiling and deep sequencing are being used to identify many biotic stress-responsive miRNAs, which could be exploited to engineer biotic stress resistance in soybean.

Conclusion

Selecting a single resistant breeding approach is challenging for breeders. In the past, judgements were made in light of data such as heredity, ease of selection, *etc.* that allowed the breeder to concentrate on creating a resistant plant phenotype. This frequently hasn't led to long-lasting resistance. An technique where the end goal is a resistant crop phenotype instead of selection for a resistant plant phenotype. This amount of plant disease will be considered commercially acceptable. Once this degree of crop resistance has been achieved, the optimum breeding method can be chosen. In order to achieve a desirable crop phenotype that includes quality and yield features, host plant resistance would need to be integrated into a crop management system. This will not only need to consider the availability of plant resistance genes (R genes and/ or QTLs), but also other aspects affecting crop phenotype, such as agronomy, breeding practises, alternative control techniques, and most critically, the pathogen's capability to evolve novel pathotypes. Even though the resistance phenotype of individual plant genotypes is not persistent, a crop management system can generate a long-lasting crop resistance phenotype by including host plant resistance.

REFERENCES

1. Food and Agriculture Organization (2019) The State of Food and Agriculture 2019: Moving Forward on Food Loss and Waste Reduction. http: //www.fao.org/state-of-food-agriculture/2019/en. Accessed 8 February 2023.
2. Savary S, Willocquet L, Pethybridge SJ, Esker P, McRoberts N, Nelson A. The global burden of pathogens and pests on major food crops. Nature Ecology Evolution. 2019. 3(3): 430–439. https: //doi.org/10.1038/s41559-018-0793-y
3. Singh B D. Plant Breeding ; Principles and Methods. Kalyani Publishers, New Delhi, 1983. 367-481.
4. Miah G., Rafii MY, Ismail MR, Puteh AB, Rahim HA, Asfaliza R. and Latif MA. Blast resistance in rice: A review of conventional breeding to molecular approaches. Molecular Biology Reports. 2013: 40: 2369–2388.
5. Anjum ZI, Hayat K, Chalkin S, Azhar TM, Shehzad U, Ashraf F, Tariq M, Mehmood HT, Azam M. Development of cotton leaf curls virus tolerance varieties through interspecific hybridization. African Journal of Agricultural Research.2015; 10(13): 1612-1618.
6. Singh P. Essentials of Plant Breeding. Kalyani Publication, 2013, 209

7. Gnanamanickam SS, Priyadarisini VB, Narayanan NN, Vasudevan P, Kavitha, S. An overview of bacterial blight disease of rice and strategies for its management. Current Science. 1999; 77: 1435–1444.
8. Khush GS. What it will take to Feed 5.0 Billion Rice consumers in 2030. Plant Molecular Biology. 2005; 59: 1–6
9. Borlaug N.E. The use of multilineal or composite varieties to control air-borne epidemic disease of self pollinated crop plants. In: Proc. First Intl. Wheat Genet. Symp. University of Manitoba, Winnipeg, 1958, 12-31
10. Bonman JM, Khush, GS, Nelson, RJ. Breeding rice for resistance to pests. Annual review of phytopathology. 1992; 30: 507–528
11. Adugna A. Alternative approaches in deploying genes for disease resistance in crop plants. Asian Journal of Plant Sciences. 2004; 3(5): 618-623.
12. Marshal, D.R. The advantages and hazards of genetic homogeneity. Annals of the New York Academy of Sciences, 1977; 287: 1-20
13. Kumar R, Mandal SS, Mishra, AK, Smriti, SR, Kumar, P. Heterosis and combining ability for yield and its contributing traits of kharif maize (*Zea mays* L.). The Bioscan. 2015. 10 (4): 2049-2056.
14. Mishra, A.K., Singh, P.K., Kumar, A., Kumar, M., Kumar, R., Singh R., Singh, S.P. Bacterial leaf blight resistance through marker assisted approach among elite rice (*Oryza sativa* L.) varieties. International Symposium on Management of rice based agricultural Breeding for System under stress prone environment. 2016.
15. Hickey LT, Germán SE, Pereyra SA, Diaz JE, Ziems LA, Fowler RA, Platz GJ, Franckowiak JD, Dieters MJ. Speed breeding for multiple disease resistance in barley. Euphytica. 2017; 213(3): 1-14.
16. Bisht, Ashita, Dinesh Kumar Saini, Baljeet Kaur, Ritu Batra, Sandeep Kaur, Ishveen Kaur, Suruchi Jindal *et al.*, "Multi-omics assisted breeding for biotic stress resistance in soybean." *Molecular Biology Reports* (2023): 1-28.
17. Tanin, Mohammad Jafar, Dinesh Kumar Saini, Karansher Singh Sandhu, Neeraj Pal, Santosh Gudi, Jyoti Chaudhary, and Achla Sharma. "Consensus genomic regions associated with multiple abiotic stress tolerance in wheat and implications for wheat breeding." *Scientific Reports*. 2022: 13680.

Chapter 16

Breeding against Abiotic Stress

S. Anand[1], M.O. Ankitha[1], S. Bhaskar Reddy[1] and Akhila Ashokan[2]*

[1]Department of Plant Breeding and Genetics,
[2]Department of Soil Science and Agricultural Chemistry,
College of Agriculture, Vellayani, Kerala Agricultural University, Kerala
**e-mail: anandsreevihar@gmail.com*

Introduction

Increasing population, increasing demand for water, flood, drought, failure to adapt to climate change, declining farm land, soil moisture, soil characteristics, deterioration in nutrient content, weed competitiveness, increasing intensity, and the frequency of biotic/abiotic stresses will amplify the challenges of achieving future food requirements [1]. Even though agricultural growth has been higher than the rate of population growth, concerns has been raised whether we will be capable of supporting its expanding population by 2030 and 2050. Food security remains a relevant and priority of many nations with special emphasis on developing countries. There is a growing concern about the future demand in the world. The global food system is experiencing an unprecedented confluence of pressures that may increase over the years 2050 [2]. The world population is estimated to reach 8.6 billion in 2030, increase to 9.8 billion in 2050 and further to 11.2 billion by 2100 [3]. This projected large world population in 2030 and 2050 discussed above have a great deal of attention due to its influence on world food demand.

Breeding and agronomic improvements have, on average, achieved a linear increase in food production globally, at an average rate of 32 million metric tons per year. However, by meet the recent 'Declaration of the World Summit on Food Security' *i.e.*, a target of 70 per cent more production by 2050, an average annual increase in food production of 44 million metric tons per year is required, representing a 38 per cent increase over historical increases in production, for next 40 years [4]. This scale of a sustained increase in global food production is unprecedented and requires sustainable mitigation of various abiotic stresses. Being complex quantitative traits, most of the abiotic stress resistance pathways are controlled by many small-effect genes, which have been difficult to take advantage of in practical breeding. The present chapter focuses on various precise breeding methods, their effectiveness, current trends, achievements and future prospects of plant breeding in counteracting various types of abiotic stresses.

Importance and impact of Abiotic Stress on Crop Plants

Natural and agricultural ecosystems are facing a harsh impact due to climate change. [5] World have experienced extreme fluctuations in weather events such as intense precipitations, flooding, heat-waves, drought periods and fluctuating freezing patterns which have strong impact on agriculture sector and world food security[6].

In majority of abiotic stresses namely drought, salinity, temperature and nutrient stress affecting crop plants, there crop responses mainly includes increase in levels of reactive oxygen species (ROS) resulting in damage of carbohydrates, proteins and DNA[7]. Accumulation of ROS and H_2O_2 under most of the abiotic stress are reported in major food crops like rice, wheat, tomato and soyabean [8, 9, 10, 11]. Plants when subjected to stress conditions, initiates production of certain enzymes and antioxidative compounds to counteract excess ROS accumulation. Out of these the major abiotic stress enzymes include ascorbate peroxidase (APX), peroxidase, catalase, superoxide dismutase (SOD) and the compounds includes glutathione, ascorbate and carotenoids [12]. Accumulation of ROS and H_2O_2 initiates molecular level responses like increase in cytosolic Ca^{2+} levels as a result of activation of Ca^{2+} channels [13]. These molecules acts as signalling factors resulting in a cascade of molecular events and activates genes for stress tolerance. Studies also show that several plant hormones like abscisic acid (ABA), ethylene, jasmonic acid and salicylic acid have promising role in mediating stress responses in roots due to drought, salinity, heat and cold stress.

Deeper insights on various abiotic stress response mechanisms and incorporation of genomic approaches for crop improvement can increase the efficiency of plant breeding strategies. Of various abiotic stresses maximum yield loss in crops throughout the world is reported to be due to adverse effect of water stress, salinity and high temperature. This chapter focuses on various methodological advancements and successful application of those plant breeding strategies in mitigation of major abiotic stress such as drought, salinity, and high temperature in crop plants.

Advanced Breeding Strategies for Abiotic Stress Tolerance

Most of the crop gene pools and elite germplasms have enough diversity for tolerance to abiotic stress but there is need for rapid introgression and crop breeding to develop abiotic stress adapted lines. Optimization of breeding process through effective screening technologies, appropriate donor selection, recycling of parents and faster generation advance can nave significant impact on varietal development process [14].

For the development of such efficient breeding pipelines a combinatorial approach involving conventional, transgenic, molecular and genomic approach is required. Most of the breeding approaches are not mutually exclusive, but are complementary under most of the breeding schemes [15]. Complex plant traits like abiotic stress are governed by the expression of multiple Quantitative Trait Loci (QTLs) instead of one or few genes. So efficient breeding strategies like genomic selection (GS) and marker assisted recurrent selection (MARS) can improve the precision level in phenotyping and reduces the time duration and breeding cycles [16].

Genomics-based approaches mainly including functional genomics, structural genomics and comparative genomics techniques can also be helpful in unravelling the molecular mechanism in various abiotic stress responses. Functional genomics including sequence or hybridization based methodologies provided insights on gene functions and the interactions between genes in regulatory networks. Sequencing-Based Approaches of functional genomics includes analysis of Expressed Sequence Tags (ESTs), quantification of abundant transcripts using Serial Analysis of Gene Expression (SAGE) [17] and Massively Parallel Signature Sequencing (MPSS) [18]. Array-based methods such as microarray hybridization technique, Targeting Induced Local Lesions IN Genomes (TILLING) for high-throughput mutant analysis [19] and EcoTILLING to identify natural polymorphisms are being effectively utilised in pathway elucidation and for developing specific breeding programs in plants for abiotic stress resistance.

Structural genomics focus on the physical structure of the genome, aiming to identify, locate, and order genomic features along chromosomes. It includes genome sequencing, genome mapping approaches and molecular marker based breeding [20]. Comparative genomics assists in gaining information by using the conservation between closely related plant species [21]. Genome zippers enabling the determination of a virtual gene order and multiple genome homology assessments can reveal species-specific genomic features that are very much mandatory for developing efficient breeding strategies to mitigate abiotic stress [22].

Breeding for Tolerance to Drought Stress

Drought is one of the major causes hampering agricultural production over the last 50 years. This will result in severe reduction of global crop yield due to water scarcity is expected in the near future. Drought has direct impacts on yield and causes various alterations in germination rates, crop biomass, root architecture, seed production and harvest index [23]. Cereals play major role in meeting the global food demand of which pearl millet (*Pennisetum glaucum* (L.) R. Br.) and upland rice (*Oryza sativa* L.) are mainly grown under hot and dry conditions, subjecting to water stress and thereby affecting the cop productivity. Productivity of rainfed rice is only half the productivity of irrigated systems [24].

In rice breeding, direct selection for improving drought tolerance was often ineffective because drought tolerance in rice show a high degrees of genotype by environment interaction.

Conventional plant breeding have not succeeded in developing high yielding varieties with improved tolerance levels to drought stress. Reason for this slowed developments in breeding programs is attributed to be due to the complex nature of drought stress in rice and its strong interaction with the environment. Other than that most of the drought stress responses are governed by quantitative trait loci (QTLs), which contribute to the final expression of various traits through interaction among themselves, their individual effects and with the environment [25].

Recent advancements in the field of genomics offer new opportunities to dissect the QTLs for drought tolerance. In rice large number of QTLs associated with drought tolerance have been mapped [26]. The major effect QTLs controlling grain yield under drought including $qDTY_{1.1}$ [27], $qDTY_{2.1,}$ $qDTY_{3.1}$ [28] and $qDTY_{12.1}$ [29] were identified to have significant role in imparting drought tolerance.

To enhance a crop's ability to avoid drought stress, the introduction of genes capable of inducing deeper root system into shallow-rooting cultivars is an effective breeding strategy. In cereals, deep rooting trait is determined by the combination of a large root growth angle (RGA; the angle between the soil surface and the shallowest primary root) and long seminal and nodal roots as RGA determines the dominant direction of root elongation [31]. Three major rice QTLs for RGA *i.e. DRO1* (*DEEPER ROOTING 1*), *DRO2* and *qSOR1* (quantitative trait locus for *SOIL SURFACE ROOTING 1*), were also reported in three different mapping populations and they can be widely used in breeding programs [31].

Barley lines that overexpressed a gene encoding for the cytokinin dehydrogenase in roots showed more tolerance to drought stress [32]. Flavonoid, phytohormone, phenolamides and antioxidant are involved in plant response to drought stress. Using transgenic technology TaPRX-2A, a wheat class III peroxidase gene family depicted a positive regulator under drought stress whose overexpression helps in enhancing drought tolerance in wheat [33].

Breeding for Tolerance to Salinity Stress

Plant responses are highly correlated with various soil parameters of which electrical conductivity defines the level of osmotic pressure and salinity in the soil. Soil is considered saline when electrical conductivity of a saturated soil extract is ≥ 4 dS/m, which is equivalent to 40 mM NaCl [34]. Alleviation of salinity stress is mainly achieved by accumulation of abscisic acid in the plant system resulting in improved photosynthesis, increased assimilate translocation and growth. Other than abscisic acid, increase in production of various biomolecules like salicylic acid, brassinosteroids, nitric oxide, polyamines and heat shock proteins [35]. Increase in antioxidant activities of enzymes and improved plant metabolic responses thereby helps the plants to cop up under salt stress conditions [36].

Genomic dimensions of these physiological responses can be exploited by breeders to develop crops tolerant to salinity stress. Mutation breeding ion barley revealed interpretative difference between the salt-tolerant genotypes *i.e.*, M4-73-30 and its wild type due to differential expression of *Hv*SOS2 (24-fold), *HvSOS3* (31-fold), *HvSOS1* (105-fold) and *HVA* (202-fold) genes in roots. Decreased sodium accumulation was observed in mutant type due to less transfer of sodium ions from root to shoot thereby providing salinity tolerance [37]. Similarly, ethyl methyl sulfonate (EMS) treatment derived tolerant mutants in wheat.

Physical and genetic maps of wild relatives of crops can also be utilized for selection of genotypes which can be used for gene transfer experiments. A halophytic relative of wild rice, *Oryza coarctata is reported to be tolerant to salt and submergence stress* [38]. Salt tolerance in 15 lines of wheat was also reported in doubled haploids of wheat [39]. Double-haploid development technology is an efficient method for fixation of favourable alleles controlling stress tolerant and agronomically important traits.

Advancement in molecular biology have made development of precise molecular markers, like single-nucleotide polymorphisms (SNPs), simple sequence repeats (SSR) and insertion-deletions (InDels) for various tolerant genes and QTLs easier. Such markers are utilized in marker-assisted breeding aiding accurate and precise selection for salinity tolerance in many of the major cereals and pulses [40,41]. Till date, use of marker assisted backcross strategy in rice was successful in introduction of a major salt tolerance governing QTL '*SalTol*' from pokkali rice to a wide range of rice genotypes [42].

Molecular marker based Genome-Wide Association Studies (GWAS) provides an insight in identification of genotype–phenotype association focusing on linkage disequilibrium and recombination, and single feature polymorphism and helps to elucidate specific genes or polymorphism within the encoding regions. GWAS mapping in rice seedling stage revealed novel 26 QTLs from 179 rice landraces genotyped with 21,623 SNP markers for salinity stress tolerance when treated with 100 mM NaCl treatment [43]. This methodology was extended to barley and a salinity tolerant QTL was identified with *DArT* marker at chromosome 4 [44].

In transgenic rice, overexpression of *Oryza sativa* trehalose phosphate synthase (*OsTPS1*) shows increased trehalose synthesis and improved salt tolerance. Similarly, upregulation of early light-inducible protein (*ELIP*), heat shock protein (*HSP70*) and water stress-inducible protein (*WSI18*) have resulted in the enhancement of salt tolerance. Similar successful introversions have been reported in many plant species namely Arabidopsis, tobacco, maize, cotton, groundnut, finger millet and rapeseed [35]. Rice salt tolerance has been linked to overexpression of the three salt overly sensitive (SOS) *loci* (*SOS1, SOS2, and SOS3*), that determines both plant salt stress tolerance and maintain a low cellular Na^+: K^+ ratio [45].

CRISPR/Cas9 is a precise, systematic, and appropriate method of genome editing which has been utilized extensively in several crops. Salt tolerance of a japonica rice cultivar WPB106 was improved by knocking out *OsRR22* gene using a *Cas9-OsRR22-gRNA* expression vector [46]. In cotton knockout of *GhFAD2* genes using CRISPR/Cas 9 aided the development of first high-oleic and non-transgenic allotetraploid upland cotton mutant [47]. Nowadays, high throughput imaging based selection methods like hyperspectral imaging has also been effectively utilized in screening salt tolerant lines in wheat crop [48].

Breeding for Tolerance to High Temperature Stress

Wheat, rice, maize, and soybean, which supply two-thirds of the human food worldwide are expected to suffer yield losses of 6.0, 3.2, 7.4, and 3.1 per cent respectively, for each degree-Celsius increase in the global mean temperature [49]. Effects linked to high temperature include changes in shoot and root growth, biomass production, flowering time, and seed production that in turn affect crop yield [50].

Heat tolerance responses is mediated by a series of complicated signaling pathways including free radical scavengers, ion transporters, signaling cascades, transcriptional and post-transcriptional control of a series of proteins. The coordinated approach involving all these components are necessary to obtain resistance to heat stress [51].

Construction of high-density QTL map for heat resistance could provide a strong foundation for molecular marker-assisted breeding. Mapping population developed using heat-resistant rice strain HT54 and the heat-sensitive strain HT13 were used to locate *OsHTAS* gene that encodes a ubiquitin ligase which regulate stomatal closure by induction of H_2O_2 under heat stress [52,53]. Other heat-resistant QTL locus, *TT1, TT2* also have broad application potential in high temperature resistance breeding programmes in grass crops including wheat, maize and chinese cabbage [51].

The application of genome-wide association studies (GWAS) permits narrowing down the candidate regions to identify specific haplotypes determining heat stress tolerance in crop plants. The method was successfully demonstrated in identifying heat stress associated key genomic regions in sorgum [54]. In cowpea QTLs governing pod set number per peduncle under heat stress is identified and utilized in breeding applications [55]. Through syntenic analysis of the closely related soybean genomes, HSPs and HSFs were also identified to regulate heat stress responses [56].

Biomolecules like melatonin interacts with other phytohormones and gaseous molecules, which greatly supports plant adaptation to temperature stress. Overexpression of *OgTT1* locus was associated with enhanced heat stress tolerance in Arabidopsis, rice, and *Festuca elata* Keng f. ex E.B.Alexeev [57].

Genetic engineering via overexpression and CRISPR/Cas system of melatonin biosynthetic genes uplifts the melatonin levels in transgenic plants and enhances temperature stress tolerance in tomato, tobacco, bermuda grass [58]. A set of four new small *Hsp26* (*sHsp26*) alleles in durum wheat suitable for enhancing heat tolerance was identified using both *in silico* and *in vivo* TILLING approaches. Such tolerant lines can be used further for breeding programs.

Conclusion and Future Prospects

Development in omics technologies have favoured the development of various crop varieties tolerant diverse stress conditions. Mostly genomics tools proved highly success in case of understanding the molecular mechanisms and identification of candidate genes for abiotic stress tolerance [59]. Despite of strenuous efforts, there is a need of development and implementation of plant breeding methods like genomic selection which can favour faster tolerant variety development with high accuracy and improved yield.

The researches in the recent past have been concentrating mainly on the effects of single abiotic stress in major food crops. Usually abiotic stress tend to occur in combinations and thereby initiating much more complex physiological, metabolic and molecular responses in crops which differ widely from that of individual stresses responses. Although studies are presently initiated in the area of combined stress in the recent past but more research should through light on various signalling mechanisms activated in the roots system [60], simultaneous signalling cascades, and expression of diverse genes and proteomics dimensions involved in abiotic stress responses. Understanding such complex regulatory mechanisms underlying plant response would help to find innovative breeding tools for developing more climate resilient crops [61].

In addition, the involvement of epigenetic mechanisms in abiotic stress tolerance in roots is still poorly known and more emphasis is to be given to decipher the underlining molecular mechanism and regulations. Development and implementation of artificial intelligence and imaging technologies in plant selection and stress tolerance screening can provide a longer leap for plant breeding in counterfeiting abiotic stress so as to achieve food security to the growing human population.

REFERENCES

1. Sandhu, N., and Kumar, A. (2017). Bridging the rice yield gaps under drought: QTLs, genes, and their use in breeding programs. *Agronomy*, *7*(2), 27.
2. Foresight. The Future of Food and Farming. Executive Summary, The Government Office for Science, London. 2011. Available from: https: // www.gov.uk/government/publications/future-of-foodand-farming.
3. UNDESA United Nations-Department of Economic and Social Affairs]. 2015. *Population Division, World Population Prospects: The 2015 Revision* (online). Available: https: //en.wikipedia. org/wiki/Projections_of_population_growth#cite_note-UN-WPP-2015-4.
4. Tester, M., and Langridge, P. (2010). Breeding technologies to increase crop production in a changing world. *Science*, *327*(5967), 818-822.
5. Watts, N., Amann, M., Arnell, N., Ayeb-Karlsson, S., Beagley, J., Belesova, K., and Costello, A. (2021). The 2020 report of the Lancet Countdown on health and climate change: responding to converging crises. *The lancet*, *397*(10269), 129-170.
6. Dempewolf, H., Eastwood, R. J., Guarino, L., Khoury, C. K., Müller, J. V., and Toll, J. (2014). Adapting agriculture to climate change: a global initiative to collect, conserve, and use crop wild relatives. *Agroecology and Sustainable Food Systems*, *38*(4), 369-377.
7. Foyer, C. H., and Noctor, G. (2009). Redox regulation in photosynthetic organisms: signaling, acclimation, and practical implications. *Antioxidants and redox signaling*, *11*(4), 861-905.
8. Zhao, C., Liu, B., Piao, S., Wang, X., Lobell, D. B., Huang, Y., and Asseng, S. (2017). Temperature increase reduces global yields of major crops in four independent estimates. *Proceedings of the National Academy of sciences*, *114*(35), 9326-9331.
9. Kamal, A. H. M., and Komatsu, S. (2015). Involvement of reactive oxygen species and mitochondrial proteins in biophoton emission in roots of soybean plants under flooding stress. *Journal of Proteome Research*, *14*(5), 2219-2236.
10. Khan, M. Y., Prakash, V., Yadav, V., Chauhan, D. K., Prasad, S. M., Ramawat, N., and Sharma, S. (2019). Regulation of cadmium toxicity in roots of tomato by indole acetic acid with special emphasis on reactive oxygen species production and their scavenging. *Plant Physiology and Biochemistry*, *142*, 193-201.

11. Kushwaha, B. K., Singh, S., Tripathi, D. K., Sharma, S., Prasad, S. M., Chauhan, D. K., and Singh, V. P. (2019). New adventitious root formation and primary root biomass accumulation are regulated by nitric oxide and reactive oxygen species in rice seedlings under arsenate stress. *Journal of hazardous materials, 361*, 134-140.

12. Desikan, R., Cheung, M. K., Bright, J., Henson, D., Hancock, J. T., and Neill, S. J. (2004). ABA, hydrogen peroxide and nitric oxide signalling in stomatal guard cells. *Journal of experimental botany, 55*(395), 205-212.

13. Ranty, B., Aldon, D., Cotelle, V., Galaud, J. P., Thuleau, P., and Mazars, C. (2016). Calcium sensors as key hubs in plant responses to biotic and abiotic stresses. *Frontiers in Plant Science, 7*, 327.

14. Mondal, S., Sallam, A., Sehgal, D., Sukumaran, S., Farhad, M., Navaneetha Krishnan, and Biswal, A. *et al.* (2021). Advances in breeding for abiotic stress tolerance in wheat. *Genomic Designing for Abiotic Stress Resistant Cereal Crops*, 71-103.

15. Ribaut, J. M., De Vicente, M. C., and Delannay, X. (2010). Molecular breeding in developing countries: challenges and perspectives. *Current opinion in plant biology, 13*(2), 213-218.

16. Gazal, A., Dar, Z. A., and Lone, A. A. (2018). Molecular breeding for abiotic stresses in maize (Zea mays L.). *Maize germplasm: Characterization and genetic approaches for crop improvement*, 25-38.

17. Vega-Sánchez, M. E., Gowda, M., and Wang, G. L. (2007). Tag-based approaches for deep transcriptome analysis in plants. *Plant Science, 173*(4), 371-380.

18. Brenner, S., Johnson, M., Bridgham, J., Golda, G., Lloyd, D. H., Johnson, D., and Corcoran, K. (2000). Gene expression analysis by massively parallel signature sequencing (MPSS) on microbead arrays. *Nature biotechnology, 18*(6), 630-634.

19. McCallum, C. M., Comai, L., Greene, E. A., and Henikoff, S. (2000). T argeting I nduced Local Lesions IN Genomes (TILLING) for plant functional genomics. *Plant physiology, 123*(2), 439-442.

20. Edwards, D., and Batley, J. (2010). Plant genome sequencing: applications for crop improvement. *Plant biotechnology journal*, 8(1), 2-9.

21. Guyot, R., Lefebvre-Pautigny, F., Tranchant-Dubreuil, C., Rigoreau, M., Hamon, P., Leroy, T., and De Kochko, A. (2012). Ancestral synteny shared between distantly-related plant species from the asterid (*Coffea canephora* and *Solanum Sp.*) and rosid (*Vitis vinifera*) clades. *BMC genomics, 13*(1), 1-14.

22. Mayer, K. F., Taudien, S., Martis, M., S¡imková, H., Suchánková, P., Gundlach, H., and Stein, N. (2009). Gene content and virtual gene order of barley chromosome 1H. *Plant Physiology, 151*(2), 496-505.

23. Dietz, K. J., Zörb, C., and Geilfus, C. M. (2021). Drought and crop yield. *Plant Biology, 23*(6), 881-893

24. Hossain, M. and Narciso, J. 2004 Global rice economy: long-term perspectives. FAO Rice Conference 04/CRS. Rome, Italy: Food and Agriculture Organization.

25. Shanmugavadivel, P. S., Prakash, C., and Mithra, S. A. (2019). Molecular approaches for dissecting and improving drought and heat tolerance in rice. In *Advances in rice research for abiotic stress tolerance* (pp. 839-867). Woodhead Publishing.

26. Ye, G., Ogbonnaya, F., and Ginkel, M. V. (2010). Marker-assisted recurrent backcrossing in cultivar development. *Molecular plant breeding: principle, method and application*, 295-319.

27. Vikram, P., Swamy, B. M., Dixit, S., Ahmed, H. U., Teresa Sta Cruz, M., Singh, A. K., and Kumar, A. (2011). qDTY 1.1, a major QTL for rice grain yield under reproductive-stage drought stress with a consistent effect in multiple elite genetic backgrounds. *BMC genetics, 12*, 1-15.

28. Venuprasad, R., Dalid, C. O., Del Valle, M., Zhao, D., Espiritu, M., Sta Cruz, M. T., and Atlin, G. N. (2009). Identification and characterization of large-effect quantitative trait loci for grain yield under lowland drought stress in rice using bulk-segregant analysis. *Theoretical and Applied Genetics, 120*, 177-190.

29. Bernier, J., Kumar, A., Ramaiah, V., Spaner, D., and Atlin, G. (2007). A large effect QTL for grain yield under reproductive stage drought stress in upland rice. *Crop Science, 47*(2), 507-516.

30. Araki, H., Morita, S., Tatsumi, J., and Iijima, M. (2002). Physiol-morphological analysis on axile root growth in upland rice. *Plant production science, 5*(4), 286-293.

31. Uga, Y., Okuno, K., and Yano, M. (2011). Dro1, a major QTL involved in deep rooting of rice under upland field conditions. *Journal of experimental botany, 62*(8), 2485-2494.

32. Pospíšilová, H., Jiskrova, E., Vojta, P., Mrizova, K., Kokáš, F., Èudejková, M. M. and Galuszka, P. (2016). Transgenic barley overexpressing a cytokinin dehydrogenase gene shows greater tolerance to drought stress. *New biotechnology, 33*(5), 692-705.

33. Yan, J., Sui, C., Liu, X., Zhang, G., Wang, S., and Guo, S. (2023). Integrated Transcriptomics and Functional Characterization Reveals that the Class III Peroxidase Gene Taprx-2a Regulates Drought Stress Tolerance in Transgenic Wheat. *eroxidase Gene Taprx-2a Regulates Drought Stress Tolerance in Transgenic Wheat.*

34. Brown Jr, G. E. (2008). Research databases. Bibliography on salt tolerance. *USDA-ARS. US Dep. Agric. Res. Serv. Riverside, CA. http: //www. ars. usda. gov/Services/docs. htm/docid, 8908.*

35. Singh, M., Nara, U., Kumar, A., Choudhary, A., Singh, H., and Thapa, S. (2021). Salinity tolerance mechanisms and their breeding implications. *Journal of Genetic Engineering and Biotechnology, 19*(1), 1-18.

36. Su, P., Yan, J., Li, W., Wang, L., Zhao, J., Ma, X.,. and Kong, L. (2020). A member of wheat class III peroxidase gene family, TaPRX-2A, enhanced the tolerance of salt stress. *BMC plant biology, 20*, 1-15.

37. Yousefirad, S., Soltanloo, H., Ramezanpour, S. S., Zaynalinezhad, K., and Shariati, V. (2018). Salt oversensitivity derived from mutation breeding improves salinity tolerance in barley via ion homeostasis. *Biologia Plantarum, 62*, 775-785.

38. Garg, Rohini, Verma, Mohit, Agrawal, Shashank, Shankar, Rama, Majee, Manoj, and Jain, Mukesh (2014). Deep transcriptome sequencing of wild halophyte rice, Porteresia coarctata, provides novel insights into the salinity and submergence tolerance factors. *DNA research, 21*(1), 69-84.

39. Bidhan, R., Noren, S. K., Mandal, A. B., and Basu, A. K. (2011). Genetic engineering for abiotic stress tolerance in agricultural crops. *Biotechnology, 10*(1), 1-22.

40. Singh, V. K., Singh, B. D., Kumar, A., Maurya, S., Krishnan, S. G., Vinod, K. K., and Singh, A. K. (2018). Marker-assisted introgression of Saltol QTL enhances seedling stage salt tolerance in the rice variety "Pusa Basmati 1". *International journal of genomics, 2018.*

41. Pushpavalli, R., Krishnamurthy, L., Thudi, M., Gaur, P. M., Rao, M. V., Siddique, K. H., and Vadez, V. (2015). Two key genomic regions harbour QTLs for salinity tolerance in ICCV 2× JG 11 derived chickpea (Cicer arietinum L.) recombinant inbred lines. *BMC plant biology, 15*(1), 1-15.

42. Gregorio, G. B., Islam, M. R., Vergara, G. V., and Thirumeni, S. (2013). Recent advances in rice science to design salinity and other abiotic stress tolerant rice varieties. *Sabrao J. Breed. Genet, 45*(1), 31-41.

43. Le, T. D., Gathignol, F., Vu, H. T., Nguyen, K. L., Tran, L. H., Vu, H. T. T.,. and Hoang, G. T. (2021). Genome-wide association mapping of salinity tolerance at the seedling stage in a panel of Vietnamese landraces reveals new valuable QTLs for salinity stress tolerance breeding in rice. *Plants, 10*(6), 1088.

44. Fan, L., Li, H., Zhuo, J., Zhang, Y., Wang, J., Chen, L., and Jiang, T. (2016). The human brainnetome atlas: a new brain atlas based on connectional architecture. *Cerebral cortex, 26*(8), 3508-3526.

45. Martínez-Atienza, J., Jiang, X., Garciadeblas, B., Mendoza, I., Zhu, J. K., Pardo, J. M., and Quintero, F. J. (2007). Conservation of the salt overly sensitive pathway in rice. *Plant physiology, 143*(2), 1001-1012.

46. Zhang, A., Liu, Y., Wang, F., Li, T., Chen, Z., Kong, D., and Luo, L. (2019). Enhanced rice salinity tolerance via CRISPR/Cas9-targeted mutagenesis of the OsRR22 gene. *Molecular breeding, 39*, 1-10.

47. Chen, Y., Fu, M., Li, H., Wang, L., Liu, R., Liu, Z., and Jin, S. (2021). High oleic acid content, nontransgenic allotetraploid cotton (Gossypium hirsutum L.) generated by knockout of GhFAD2 genes with CRISPR/Cas9 system. *Plant biotechnology journal, 19*(3), 424.

48. Moghimi, A., Yang, C., Miller, M. E., Kianian, S. F., and Marchetto, P. M. (2018). A novel approach to assess salt stress tolerance in wheat using hyperspectral imaging. *Frontiers in plant science, 9*, 1182.

49. Zhao, Y., Li, Y., and Yang, F. (2021). Critical review on soil phosphorus migration and transformation under freezing-thawing cycles and typical regulatory measurements. *Science of the Total Environment, 751*, 141614.

50. Li, T., Hasegawa, T., Yin, X., Zhu, Y., Boote, K., Adam, M., and Bouman, B. (2015). Uncertainties in predicting rice yield by current crop models under a wide range of climatic conditions. *Global change biology, 21*(3), 1328-1341.

51. Yang, Y., Yu, J., Qian, Q., and Shang, L. (2022). Enhancement of Heat and Drought Stress Tolerance in Rice by Genetic Manipulation: A Systematic Review. *Rice, 15*(1), 1-18.

52. Wei, H., Liu, J., Wang, Y., Huang, N., Zhang, X., Wang, L., and Zhong, X. (2013). A dominant major locus in chromosome 9 of rice (*Oryza sativa* L.) confers tolerance to 48 C high temperature at seedling stage. *Journal of Heredity, 104*(2), 287-294.

53. Liu, J., Zhang, C., Wei, C., Liu, X., Wang, M., Yu, F., and Tu, J. (2016). The RING finger ubiquitin E3 ligase OsHTAS enhances heat tolerance by promoting H2O2-induced stomatal closure in rice. *Plant physiology, 170*(1), 429-443.

54. Chopra, R., Burow, G., Burke, J. J., Gladman, N., and Xin, Z. (2017). Genome-wide association analysis of seedling traits in diverse Sorghum germplasm under thermal stress. *BMC plant biology, 17*, 1-15.

55. Pottorff, M., Roberts, P. A., Close, T. J., Lonardi, S., Wanamaker, S., and Ehlers, J. D. (2014). Identification of candidate genes and molecular markers for heat-induced brown discoloration of seed coats in cowpea Vigna unguiculata (L.) Walp]. *BMC genomics, 15*, 1-11.

56. Liu, Y., Li, J., Zhu, Y., Jones, A., Rose, R. J., and Song, Y. (2019). Heat stress in legume seed setting: effects, causes, and future prospects. *Frontiers in Plant Science, 10*, 938.

57. Li, X. M., Chao, D. Y., Wu, Y., Huang, X., Chen, K., Cui, L. G.,. and Lin, H. X. (2015). Natural alleles of a proteasome α2 subunit gene contribute to thermotolerance and adaptation of African rice. *Nature genetics, 47*(7), 827-833.

58. Raza, A., Charagh, S., García-Caparrós, P., Rahman, M. A., Ogwugwa, V. H., Saeed, F., and Jin, W. (2022). Melatonin-mediated temperature stress tolerance in plants. *GM Crops and Food, 13*(1), 196-217.

59. Akpýnar, B. A., Lucas, S. J., and Budak, H. (2013). Genomics approaches for crop improvement against abiotic stress. *The Scientific World Journal, 2013.*

60. Sanchez Bermudez, M., Del Pozo, J. C., and Pernas, M. (2022). Effects of Combined Abiotic Stresses Related to Climate Change on Root Growth in Crops. *Frontiers in Plant Science,* 2087.

61. Zhang, M., Gao, C., Xu, L., Niu, H., Liu, Q., Huang, Y., and Li, M. (2022). Melatonin and Indole-3-acetic acid synergistically regulate plant growth and stress resistance. *Cells, 11*(20), 3250.

Chapter 17

Biofortification as a Strategy to Fight Malnutrition

D. Dinesh Varma and M. Vennela*

Department of Genetics and Plant Breeding, Dr. Rajendra Prasad Central Agricultural University, Pusa, Bihar
****e-mail: dineshdamodaram96@gmail.com***

Introduction

Malnutrition and food insecurity in underdeveloped countries are problems that worry the entire world. According to Ruane (2010), there are 963 million hungry people in the globe. At that period, a 50 per cent increase in production will be required to meet the world's food needs. Numerous functions of micronutrients are involved in homeostasis and metabolism. Due to insufficient availability of vitamins and other micronutrients that have an impact on human health, micronutrient deficiencies cause a global issue of population malnutrition. Half of the world's population suffer from a different kind of dietary deficiency, with developing nations bearing the brunt of the problem (Ortiz-Monasterio *et al.*, 2007). Currently, 3 billion people worldwide are suffering from micronutrient malnutrition brought on by Zn and Fe deficiency, which has major health consequences. Diets oriented on staple foods that are deficient in nutrition, such as Zn and Fe, have been linked to an increased proportion of Zn and Fe deficits.

A diet that does not include an adequate quantity of one or more nutrients can lead to malnutrition. It causes cognitive impairment, developmental delays, impairments, lowered immunity, and a variety of illnesses. Malnutrition, according to the WHO, refers to deficiencies, excesses, or imbalances in the consumption of energy, protein, and other nutrients. It encompasses both undernutrition and overnutrition. Between crop kinds, there are major differences in the quantity and makeup of micronutrients. Cereal grains, roots, and tubers are significant staple foods worldwide. Despite being high in carbs, these crops may be very low in or of poor quality in terms of proteins and micronutrients. Approximately 800 million people are estimated to be malnourished worldwide, with nearly all of them living in developing nations (Sinha *et al.*, 2019).

For the most part, modern agriculture has been successful in supplying the energy demands of underprivileged populations in emerging nations. The rise in cereal production has been the focal point of agricultural research in developing nations for the past years. Agriculture, however, must now concentrate on a framework that will not simply produce more food but also offer food of higher quality. Biofortification offers a practical way to reach rural populations that are undernourished and may have little access to commercially available fortified foods and supplements. Biofortification is the combination of Latin word "fortificare" means "to make strong," and the Greek word "Bios" means "life." It involves using agronomic techniques, conventional breeding, or genetic modification to increase the bioavailable concentration of an element in agricultural plants' edible parts. The main strategy for enhancing basic food crops with iron, zinc, and provitamin A carotenoids has been conventional breeding. Creating biofortified crops also increases their ability to grow on soils with mineral compositions that are depleted or unavailable.

To ensure the success of biofortification, the following three conditions must be met:

- ✰ It must provide a high yield and be profitable for the farmer.
- ✰ It must be proven to be successful in decreasing micronutrient deficiencies in target populations and
- ✰ It must be accepted by farmers and consumers in the target regions (Hotz and McClafferty 2007).

Purpose of Biofortification

Below is a list of some of the major goals of biofortification:

- ✰ To address the challenge of inadequate dietary nutrition. Even those who lack access to a varied and nutrient-rich diet can benefit from biofortification. They typically eat what they grow for their main diet. As a result, they tend to experience nutritional deficiencies.
- ✰ By mixing more micronutrients with the ideal agronomic practices, biofortification has been shown to address the issue of malnutrition.

- ☆ They can therefore outperform the variety that farmers grow organically. Therefore, consuming biofortified crops can long-term assure a more balanced diet.
- ☆ Biofortification has been essential in enhancing food nutrition and as a result, human health in general.

Methods of Biofortification

There are three strategies that help in achieving the biofortification process:

1. Agronomic Techniques

Agronomic methods for biofortification are cheap and simple, but they must be done with great care due to the application method, kind, and environmental considerations. This strategy emphasises improved nutrient availability to plants, their efficient usage, and plant mobility as well as an increase in microbial activity for these purposes. This technique makes use of fertilisers to increase the amount of macronutrients in plants by using agronomic procedures. Crops also require other micro-minerals as iodine, copper, zinc, iron, nickel, manganese, molybdenum, *etc.* The majority of these micronutrients are easily obtainable in soil, are taken up by plants, and eventually enter the food chain. But occasionally they are applied as mineral nutrition since plants cannot absorb them.

In addition to fertilizers, nutrient mobility from the soil to edible sections of plants can be improved by using soil microbes that promote plant growth. To boost the phyto-availability of mineral elements, soil microbes from the genera Bacillus, Pseudomonas, Rhizobium, and others can be used. When there is a shortage of nitrogen, N2-fixing bacteria are crucial for boosting crop productivity. Numerous crops have mycorrhizal fungi attached to them, which can emit organic acids, siderophores, and enzymes that can break down organic molecules and raise the mineral content of edible product. Using agronomic biofortification techniques, various crops have been modified. A few examples are provided here (Monika *et al.*, 2018).

- ☆ CEREALS: Iron was effectively increased in rice grains by biofortifying rice plants with iron-foliar spray. However, adding zinc fertiliser to the soil in addition to foliar spraying has proven to be a valuable tactic for increasing the grain zinc content of rice cultivated in soils with low background levels of zinc. Increased zinc content in grains due to the use of zinc-containing fertilisers undoubtedly improved human nutrition and health in Turkey, particularly in rural areas where more than half of daily calories were obtained from wheat. The use of several organic and inorganic biofertilizers has enhanced the micronutrient profile of barley.
- ☆ LEGUMES: By applying selenium complex salts to the leaves of soybean plants as fertiliser, selenium-enriched soybean has been created. By employing actinobacteria that promote plant growth, the

mineral limitations of chickpea have been targeted, particularly those of iron, zinc, calcium, copper, manganese, and magnesium.

- OILSEEDS: Azospirillum brasilense and Azotobacter vinelandii, two plant growth-promoting rhizobacteria, were added to canola together with chemical fertilisers, that resulted in an increase of seed's protein, oleic acid, and linoleic acid content, demonstrating that rhizobacteria are highly successful at boosting canola oil's yield and nutritional value.
- VEGETABLES AND FRUITS: Using foliar zinc fertilisers, field tests were conducted to boost the zinc concentrations in potato tubers, which resulted in a notable rise in tuber zinc concentrations. With irrigation and chemical fertiliser treatments, there has been evidence of an increase in beta-carotene in orange-fleshed sweet potatoes. KIO3 and Na2SeO4 were used as a foliar spray and nutrient medium to achieve lettuce I and Se biofortification. According to studies, tomatoes are a great crop for iodine biofortification programmes when fertilised with iron.

2. Plant Selective Breeding

This method makes use of conventional plant breeding practices. For a given attribute that is wanted, there are enough genetic variations produced. It is done to cross different plant breeds over several generations. Finally, a plant with lots of nutrients is produced. In India, this is the only method that is used. Conventional breeding, aside from agronomic biofortification, depends on the genetic diversity for the trait of interest (TOI) in the gene pool. Using traditional crossing approaches, the desired genes are pyramided, and then segregation populations are thoroughly screened. In this method, recipient lines with desirable agronomic qualities are crossed with parent lines with high nutritional content over several generations to generate plants with the desired nutrient and agronomic traits. However, the limited genetic diversity that exists in the gene pool must occasionally be taken into account in breeding methods. This can sometimes be avoided by crossing two distant relatives, which gradually introduces the trait into the commercial cultivars. Alternately, mutagenesis can be used to add novel features directly into commercial types. In India, through conventional breeding many biofortified crops are developed some of the examples include:

- A pure line variety of rice (CR Dhan 310) contains 10.3 per cent protein in polished grain as compared to 7-8 per cent of popular varieties and other variety DRR Dhan 45 contain high zinc content of 22.6 ppm was developed.
- Wheat variety WB 02 is Rich in zinc (42.0 ppm) and iron (40.0 ppm) in comparison to 32.0 ppm zinc and 28.0-32.0 ppm iron in popular varieties.

- ☆ First provitamin-A rich maize hybrid Pusa Vivek QPM9 Improved is produced in India which contains provitamin- A (8.15 ppm), lysine (2.67 per cent) and tryptophan (0.74 per cent).
- ☆ In pearl millet, hybrid Hhb 299 is developed that contains high amounts of iron (73 ppm) and zinc (41 ppm).
- ☆ Lentil: Pusa Ageti Masoor variety contains 65.0 ppm iron as compared to 55.0 ppm iron in popular varieties.
- ☆ Mustard: Pusa Mustard 30, contains low erucic acid (<2 per cent) in oil as compared to >40 per cent erucic acid in popular varieties.
- ☆ Pusa Beta Kesari 1, cauliflower variety is country's first biofortified cauliflower that contains high β-carotene (8.0-10.0 ppm) in comparison to negligible β-carotene content in popular varieties.
- ☆ Potato: Bhu Sona- High β-carotene (14.0 mg/100 g) content as compared to 2.0-3.0 mg/100 g β- carotene in popular varieties

HarvestPlus Programme

The HarvestPlus programme, developed by the CGIAR in collaboration with the International Center for Tropical Agriculture (CIAT) and the International Food Policy Research Institute, aims to breed staple foods that are biofortified. It works to eradicate hidden hunger and micronutrient deficiency around the world through agriculture and nutrition. It focuses on the most at-risk communities and offers them a food-based approach to managing micronutrient deficiencies. The company is collaborating with its partners to create crop types that are high in iron, zinc, and vitamin A, the three nutrients that the World Health Organization has recognised as being most missing in the world's foods. Since the launch of HarvestPlus, 242 crop varieties have been made available in 30 different nations. 8.5 million smallholder farming communities are cultivating these varieties and supplying them to 42.4 million people worldwide.

3. New Breeding Techniques

This method involves introducing a desired trait into an organism's DNA genome. By creating new possibilities for the development of novel genetic variation that does not already exist in the gene pool, emerging breeding techniques like as transgenic breeding, RNA interference, and genome editing are playing major roles in the biofortification of crops.

Transgenic Breeding

- ☆ Rice: With its enormous potential to minimise disease burden by expressing the genes encoding PSY and carotene desaturase, golden rice represented a substantial advancement in this direction as a reliable supply of beta-carotene. By focusing on the gene encoding carotene desaturase, phytoene, a precursor to beta-carotene, has seen an up to 23-fold increase in concentration. By overexpressing the genes for the antioxidants aminodeoxychorismate synthase and Arabidopsis GTP-cyclohydrolase I (GTPCHI), rice has been genetically altered to boost its folate content (up to 150-fold).

- Wheat: The introduction and overexpression of the genes corresponding to micronutrients like zinc, iron, vitamin A, and proteins could be used to make up for the deficits of these micronutrients in cereal grains. Ectopic overexpression of the TaFer1-A and OsNAS2 genes resulted in substantial iron accumulation in wheat grain, increasing iron content to 80 and 85 g/g in grains, respectively. However, TaFer1-A expression varies between generations. While silencing of the phyA and TaABC13 transporter genes reduced phytic acid production (18–19 per cent) and boosted iron and zinc bioavailability, overexpression of the phytochrome (PhyA) gene increased phytase production and anti-nutrient activity.
- Maize: By bacterial crtI and crtB gene overexpression under the direction of an endosperm-specific super gema zein promotor, vitamin A accretion in maize endosperm was enhanced to as high as 9.8 g/g of DW. Similar to this, maize Psy1 gene overexpression boosted-carotene accumulation in maize. The dehydroascorbate reductase gene was overexpressed, which converted ascorbic acid from its oxidising form to its reducing form, increasing the amount of vitamin C in maize. The concentrations of tocopherol and tocotrienol in maize were increased by the overexpression of the Homogentisic acid Geranyl Geranyl Transferase (HGGT) gene, and as a result, biofortified maize was produced.

RNA Interference

RNA interference (RNAi) is a sequence-specific gene regulatory process that is fueled by dsRNA molecule that inhibits the transcription or translation of a specific gene. It is frequently used to improve crops' nutritional quality and get rid of pollutants and food allergies. Due to its capacity to chelate micronutrients and limit the bioavailability of critical elements, phytic acid (PA) is regarded as a primary anti-nutrient in cereals and pulses. An essential gene in the PA biosynthesis pathway in wheat, inositol pentakisphosphate kinase (TaIPK1), was disrupted by RNA interference (RNAi). The resulting homozygous transgenic lines of wheat exhibit reduced TaIPK1 expression, decreased PA buildup (26–58 per cent), and elevated grain phosphate, zinc, and iron levels at the T4 stage. In rice, AK and DHDPS enzymes were suppressed using RNA interference (RNAi), and as a result, mutant lines accumulated 6.6- and 21.7-fold more lysine, respectively. Similar to this, beta-carotene promotes iron and zinc absorption in lentils, peas, and chickpeas. Thus, genes that encode these substances could be overexpressed for improved bioavailability of micronutrients (Rahil *et al.*, 2021).

Genome Editing

In order to produce plants free of transgenes, plant genome editing (GE) employs sequence-specific nucleases (SSNs) for targeted and stably inherited gene alteration. It includes different types, like as TALENs, ZFNs, and the CRISPR-Cas system.

Cereals: Through genome editing, the OsVIT2 gene in rice was silenced to increase the availability of Fe. In the rice line Kitaake, the Ctrl and PSY genes from maize were specifically inserted using CRISPR-Cas9 under the control of an endosperm-specific promotor. In order to reduce AC in rice, multiplex genome editing was used to target the OsWaxy gene at three locations. 14 per cent less accumulation was seen in mutants compared to wild type. Wheat's zinc content was raised by using genome editing by specifically mutating the TaVIT2 gene. Through genome editing, Zhao *et al.* (2019) specifically targeted the bioavailability of lysine, vitamin A, iron, and zinc in sorghum. Therefore, this method makes it possible to generate multiple nutrient-dense crop cultivars in a single phase while also opening up new vistas. By developing tactics to jointly target several micronutrients while taking into account their stability, metabolic engineering applications will make significant progress in this direction.

Pulses: Examples illustrating the importance of genome editing for the biofortification of pulses are currently lacking in the literature. However, there is potential for iron and zinc biofortification through the use of genome editing. Targeting anti-nutrient genes that are responsible for the decreased bioavailability of micronutrients is another strategy used for the biofortification of pulses. Saponins are the anti-nutrient substances that, when consumed in smaller amounts, are advantageous but, when consumed in larger amounts, can act as anti-nutrients. According to research conducted using the Arabidopsis thaliana plant, the synthesis of saponins is mediated by 112 uridine diphosphate glycosyltransferases (UGTs), 246 P450 genes, and 13 OSC genes. To limit the production of saponins, it is necessary to identify and remove key regulatory genes in the pathway that produces saponins.

Troubleshooting with Genetically Modified Foods

Many underdeveloped, underprivileged countries have a history of rejecting genetically modified food. Many of them are unconvinced of the advantages and safety of biotechnology. Many developing nations have relatively little knowledge about biotechnology, and discussions of its benefits and drawbacks are almost no-existent.

- Risk of uncontrolled hybridization with conventional varieties.
- Overconsumption of certain fortified crops and natural sources of the nutrients may have harmful effects.
- Alteration to the food's nutritional value.
- GM foods potential for toxicity and allergenicity.

- Potential development of new poisons.
- Agricultural genetic diversity is at risk.
- Concerns of organic and conventional farmers.

Role of Biofortified Crops Cultivars on Reducing Human Malnutrition

Compared to non-biofortified crops, biofortified crops are more nutrient-dense, and since micronutrient retention during preparation, processing, and storage is assumed to be identical, eating biofortified staple crops increases dietary intake of all micronutrients. Over 20 million people throughout the world currently consume food from biofortified crops (Bouis and Saltzman, 2017). The Fe content in potentially pregnant women in the Philippines have increased as a result of the intake of Fe-enriched biofortified crops like rice (Haas *et al.*, 2005). In order to combat Fe deficiency, Fe-biofortified pearl millet was given to Indian schoolchildren (Finkelstein *et al.*, 2015) and Fe-biofortified beans were given to Rwandan women. Because the pro-vitamin A carotenoids are first absorbed by the body and then transformed into vitamin A's active form in response to the body's needs, it was determined that interpreting the pro-vitamin A effect was more challenging. The benefits of QPM have been shown to be widespread since children who consumed a diet of QPM maize had fewer sick days than those who consumed regular maize. When compared to a control group that simply received normal maize, young children who consumed QPM gained weight and height by 12 per cent and 9 per cent, respectively and 100 g of QPM was adequate to satisfy a child's lysine needs (Nuss *et al.*, 2011). The creation and promotion of biofortified crop cultivars will help to accomplish the SDGs by eradicating malnutrition, as all the research offer significant evidence that intake of biofortified crops improves human health status.

Climate Change and Biofortification

Climate change might undo recent progress in lowering malnutrition, and up to 1 billion people could suffer as a result (WFP, 2009). The uptake of micronutrients in common bean, field pea, lentils, and chickpea is significantly regulated by the environment for instance, the Fe content is influenced by genotype and environment interactions (Suwarto, 2011). It is clear that global warming and shifting precipitation patterns will have a negative impact on agricultural yields and that this will have a negative impact on biofortification initiatives. The number of malnourished children and per capita calorie consumption were used by the International Food Policy Research Institute (IFPRI) to assess the effects of climate change on food security and human prosperity. Without climate change, it is predicted that by 2050, the availability of calories will decline globally, increasing child malnutrition by 20 per cent ; however, with climate change, levels of child malnutrition may rise significantly (Abdul *et al.*, 2018). A warming globe would have a severe impact on crop output, which would increase starvation and make people more prone to infectious diseases. Therefore, in order to maintain sufficient micronutrient concentrations under climate change, it is necessary to create more effective cultivars.

Biofortification: Strategic Benefits

- The biofortification plan aims to benefit from the regular daily ingestion of substantial amounts of food staples by every member of the family, especially women and children who are particularly at threat for micronutrient malnutrition.
- Recurrent costs are modest after the initial expenditure is made to create seeds that can bolster themselves, and germplasm may be transferred internationally.
- The multiplier impact of plant breeding over time and space is what makes it so economically feasible.
- Furthermore, biofortification offers a realistic method of supplying naturally fortified meals to those who have limited access to commercially marketed fortified foods, which are more easily accessible in metropolitan areas to undernourished communities in relatively remote rural locations.
- Additionally, enhanced or fortified seeds have greater plant vigour, seedling viability, quicker initial emergence, and higher grain output.

Limitations of Biofortification

- This method only works with certain minerals and particular plant species since it is not often feasible to target the micronutrient into edible plant sections like seed or fruit and can occasionally lead to the build-up of desirable nutrients in the leaves or other non-edible regions of plants.
- Another significant issue is phytic acid, an antinutrient molecule, which inhibits the bioavailability of minerals.
- Although the transgenic-based technique receives far more research funding than breeding-based approaches, its success rate in terms of cultivar release is quite low since it takes a long time from target trait and gene identification to understanding the potential impact on other living forms.
- To make biofortification solutions as effective as possible, postharvest processing of each crop must be taken into account as, milling or polishing cereal seeds can remove significant amounts of minerals from the diet; the magnitude of these losses depends on genotype.
- Improvements in a crop's capacity to maintain yields with decreasing water supply and quality will also be crucial in the context of global environmental change.

Conclusion

Both malnutrition and hidden hunger exist in both wealthy and developing nations and have catastrophic impacts on the world. The recent effects of the worldwide pandemic have demonstrated the necessity for food systems to be modified to progress global improvements that can prevent food supply shortages. Additionally, climate change estimates indicate that developing countries would experience greater inequality and poverty, necessitating an increase in the nutritional value of PBFs. The most sustainable and economical approach to addressing malnutrition is biofortification. The biofortification strategy aims to benefit from all family members' regular daily ingestion of substantial amounts of food staples, especially women and children who are particularly at risk for micronutrient deficiency.

Promoting the development, manufacturing, and consumption of biofortified crops is necessary to prevent a number of health problems. Release of GM crops that address widely widespread vitamin and mineral deficiencies has been delayed due to the regulatory environment currently in place and anti-GMO lobbying campaigns. Aside from techniques to increase the nutritional content of crops, one of the main obstacles to biofortification is public acceptance. Plant breeders need to be conscious of both the potential for future advances in nutrition and health from plant breeding as well as the significant impact that agricultural research may have had in the past on nutrient consumption. Since only a small number of cultivars are finally made available to consumers, additional education and marketing efforts should be made, especially for transgenic approaches.

REFERENCES

Abdul, W., Muhammad, F., Khurram, B and Levent, O. 2018. micronutrient malnutrition and biofortification: Recent advances and future perspectives. *Plant Micronutrient Use Efficiency*. 1: 225-243.

Bouis, H.E and Saltzman, A. 2017. Improving nutrition through biofortification: A review of evidence from HarvestPlus, 2003 through 2016. *Global Food Security*. 12: 49-58.

Finkelstein, J.L., Mehta, S., Udipi, S.A., Ghugre, P.S., Luna, S.V., Wenger, M., Murray-Kolb, L.E., Przybyszewski, E.M and Haas, J.D. 2015. A Randomized Trial of Iron-Biofortified Pearl Millet in School Children in India. *Journal of Nutrition*. 145: 1576-1581.

Haas, J.D., Beard, J.L., Murray-Kolb, L.E., Del Mundo, A.M., Felix, A and Gregorio, G.B. 2005. Iron-Biofortified Rice Improves the Iron Stores of Nonanemic Filipino Women. *Journal of Nutrition*. 135: 2823-2830.

Hotz, C and McClafferty, B. 2007. From harvest to health: Challenges for developing biofortified staple foods and determining their impact on micronutrient status. *Food and Nutrition Bulletin*. 28 (2): 271-279.

Monika, G., Natasha, S., Saloni, S., Payal, K., Aman, K., Venkatesh, C and Priya, A. 2018. Biofortified crops generated by breeding, agronomy, and transgenic approaches are improving lives of millions of people around the world. *Frontiers in Nutrition*. 5: 12.

Nuss, E.T., Tanumihardjo, S.A., Fretham, S.J.B., Carlson, E.S and Georgieff, M.K. 2011. Quality Protein Maize for Africa: Closing the Protein Inadequacy Gap in Vulnerable Populations. *Advances in Nutrition*. 2: 217-224.

Ortiz-Monasterio, J.I., Palacios-Rojas, N and Meng, E. 2007. Enhancing the mineral and vitamin content of wheat and maize through plant breeding. *Journal of Cereal Science*. 46 (3): 293-307.

Rahil, S., Shakra, Jamil., Shakeel, A., Amina, N., Sipper, K., Zarmaha, A., Shamsa, K., Hafiz, M., Usman, A., Rafaqat, A and Weijun, Z. 2021. Biofortification of cereals and pulses using new breeding techniques: current and future perspectives. *Frontiers in Nutrition*. 8: 721-728.

Ruane, S. 2010. Agricultural biotechnologies in developing countries: Options and opportunities in crops, forestry, livestock, fisheries and agro-industry to face the challenges of food insecurity and climate change (ABDC-10). Guadalajara, Mexico, 1-4 March, 2010. Background document to the FAO e-mail conference on "Learning from the past: Successes and failures with agricultural biotechnologies in developing countries over the last 20 years. pp.1-18.

Sinha, P., Davis, J., Saag, L., Wanke, C., Salgame, P., Mesick, J., Horsburgh, C.R and Hochberg, N.S. 2019. Undernutrition and tuberculosis: public health implications. *International Journal of Infectious Diseases*. 219 (9): 1356-1363.

Suwarto, N., 2011. Genotype environment interaction for iron concentration of rice in central Java of Indonesia. *Rice Science*. 18: 75-78.

WFP, 2009. Emergency Food Security Assessment Handbook. Rome, WFP.

Zhao, Z.Y., Che, P., Glassman, K and Albertsen, M. 2019. Nutritionally enhanced sorghum for the arid and semiarid tropical areas of Africa. In Sorghum; Humana Press: New York, NY, USA, pp. 197–207.

Chapter 18

The Regulatory System in Plant Genetic Resource Management

Ragulakollu Sravanthi*, V. Preetikumari and Priya Rawat

Ph.D. Scholar, Genetics and Plant Breeding, Tamil Nadu Agricultural University, Coimbatore, Tamil Nadu
****e-mail: sravrvk@gmail.com***

Introduction

Plant breeding is an evolution process which involves various methods and procedure for the development of new plant varieties with desirable characters and adaptability. Thus, the process of evolution is not only effective but also enhances variation by accumulating a variety of crop types that serve as a base for future advancement. The genes required for crop improvement are present in different crop species which are contributed by their wild relatives.

Germplasm or genetic resources may be defined as the sum total of the genetic information present in the crop species and its wild relatives. Sir Otto Frankel coined the term "**genetic resources**" in 1968 demonstrated that the plant breeders being aware of this gradual loss of germplasm, failed to recognize the urgency of protecting the crop genetic resources until there was no return.

The epidemic of southern corn leaf blight in USA [1970] became the basis of international concern about the loss of genetic variation as well as initiated action to collect, conserve, evaluate and utilize plant genetic resources effectively. The several international and national level organizations, conferences and publications devoted exclusively to the plant genetic resources have been established over the past thirty years.

Collection, conservation and utilization of plant genetic resources is very essential to meet the demand of the burgeoning population as well as their uses as base material to breeders for further improvement of the crops. All the activities associated with plant genetic resources starting from collection to sustainable utilisation should be under the control of regulatory system. Regulation of PGR is exercised at international as well as national level by organisation which exclusively engaged in PGR related activities. Regulatory systems play crucial role in meeting the demand of population by assuring the conservation and utilisation of PGR for food and agriculture.

Prior to 1993, PGR were used as global genetic stocks that is available to use to anyone without any regulatory boundaries. As green revolution has displaced crop genetic diversity required for incessant progression in plant breeding. During 1980s large public ex-situ collection was created in concerned with green revolution led to establishment of several International and National organisations to look after PGR regulation. In context to plant breeding, new varieties and traits developed by breeders using germplasm or wild relatives as parental material were protected under IPR (Intellectual Property Rights). Likewise, to regulate the access to use of PGR new policy framework were initiated.

In India, ICAR-NBPGR is the apex organisation involved in the management of PGR at national level under the support of ICAR (Indian Council of Agricultural Research), New Delhi. Plant genetic resources are the heritage of natural environment. India is one of the gene-rich country contribute maximum crop diversity in world scenario but at the same time protection of PGR is very challenging while framing mutually beneficial strategies for exchange of germplasm with other countries. The organisational activities of PGR include exploration, collection, exchange, characterization, evaluation, conservation and documentation. In addition to this quarantine of all the imported PGR for research purposes is undertaken by the Bureau.

Plant Genetic Resource Management

The plant genetic resources (PGR) as components of agrobiodiversity not only cover the whole gamut of genetic resources (from advanced cultivars to primitive landraces, domesticates, semi-domesticates, wild and weedy relatives) but also the diversity of ecosystems and agroecosystems within landscapes that are exploited in some way for agriculture and forestry, and the complex set of human interactions.

The Indian PGR Management system led by the National Bureau of Plant Genetic Resources (ICAR) in partnership with commodity based ICAR institutions and SAUs, collectively known as the National Active Germplasm

Sites (NAGS) has been envisioning the challenges of PGR management at national level in the past. Equally important is an attempt to foresee the key challenges and opportunities in PGR management sector in coming decades and accordingly develop strategy and roadmap to articulate the future role of the PGR management in support of Indian agriculture. The present compilation describes the chronological resume of NBPGR, significant accomplishments in genetic and genomic resource management, and future key challenges in PGR and genomic resource management.

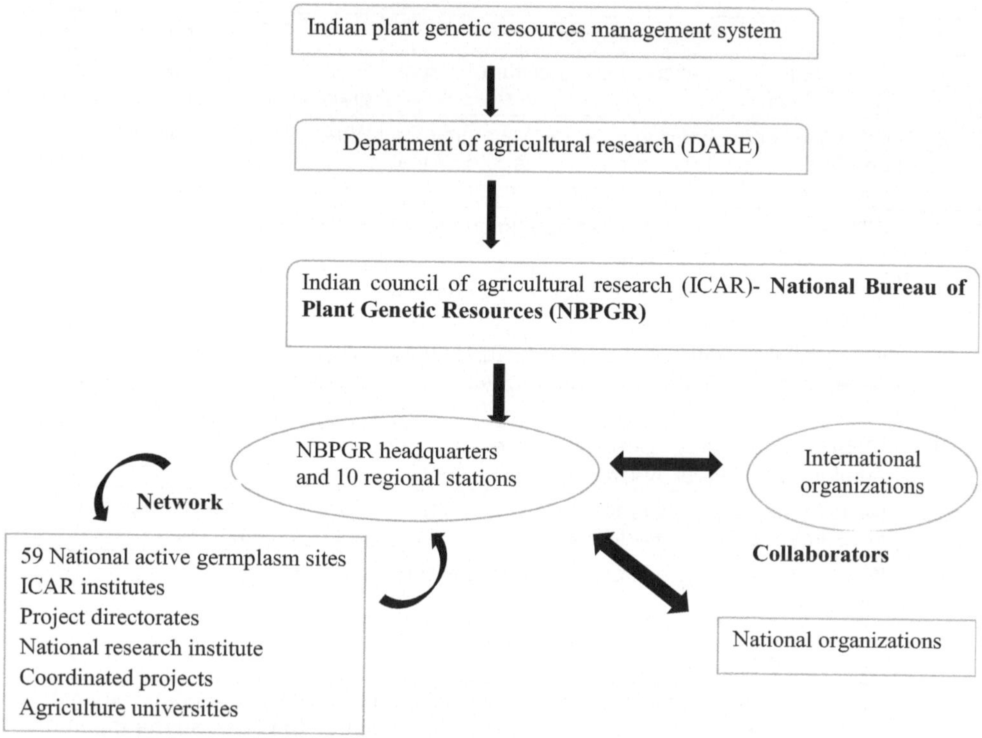

Figure 18.1: Schematic Representation of the Plant Genetic Resources Management System in India.

Exploration

Exploration provides the way to cover the areas where maximum crop diversity is found. Exploration is the primary source of germplasm collection which consists of the series of planned activities. Before exploration it is necessary to get pre information about the diversity of flora and fauna of particular geographical region. Collection of germplasm as base material for development of commercial population by breeders and collection of variability present in crop species and their wild relatives for conservation or further improvement are two basic objectives of exploration.

Collection

The collection of large number of lines, varieties and related wild species of crop is known as **Gene banks or Germplasm Collection.**

Word collection is defined as the germplasm collection is large to include entries or accessions from all over the world.

Table 18.1: Germplasm Collections in World as well as International Agricultural Research Centres

Sl.No.	*World Collection*	*Crop*
1.	CIAT-Centro Internacional De Agricultura Tropical-Cali, Colombia	Rice, forage grasses and legumes, grain legumes, cassava
2.	CIMMYT-Centro Internacional de Mejoramiento de Mais y Trigo (International Maize and Wheat Improvement Center) - El Batan, Mexico	Wheat (bread and durum), maize
3.	CIP-Centro Interactional de la Papa (International Potato Center)-Lima, Peru	Potato, sweet potato
4.	IBPGR-International Board for Plant Genetic Resources - Rome, Italy	
5.	ICARDA- International Centre for Agricultural Research in the Dry Areas - Aleppo, Syria	Wheat (bread and durum), barley, faba bean, lentil, chickpea, pigeon pea, forage grasses and legumes
6.	ICRAF-International Council for Research in Agroforestry- Nairobi, Kenya	
7.	ICRISAT-International Crops Research Institute for the Semi-Arid Tropics-Hyderabad, India	Sorghum, millet, chickpea, pigeon pea, groundnut
8.	IFPRI-International Food Policy Research Institute-Washington, D.C.	
9.	IIMI-International Irrigation Management Institute-Colombo, Sri Lanka	
10.	IITA-International Institute of Tropical Agriculture-Ibadan, Nigeria	Rice, Potato, sweet potato, ground nut, cowpea, soybean, maize, cassava
11.	ILCA- International Livestock Center for Africa – Addis Ababa, Ethiopia	
12.	ILRAD-International Laboratory for Research on Animal Diseases- Nairobi, Kenya	
13.	INIBAP-International Network for the Improvement of Banana and Plantain - Montferrier-sur-Lez, France	Banana, plantain
14.	IRRI- International Rice Research Institute - Los Banos, Philippines	Rice
15.	ISNAR-International Service for National Agricultural Research- The Hague, Netherlands	
16.	WARDA-West Africa Rice Development Association - Bouake, Cote d'Ivoire	

Evaluation

Evaluation is a comprehensive description of the collected material. Germplasm material can be provided based on the objective of the breeder as only small amount of material is relevant to specific breeding programme. A detailed information from the evaluated germplasm is made available to the breeder in selecting the breeding material.

Conservation and Management of Plant Genetic Resources

For plant improvement, most of the plant varieties and hybrids share common parents as desired trait is limited to particular parent due to which the commercial plant population faces problems like genetic variability and narrow genetic base (genetic erosion). To avoid this problem conservation of local landraces, wild relatives and exotic as well as indigenous germplasm is foremost important way to preserve diversity and utilise this variability for development of commercial varieties or hybrids with broad genetic base.

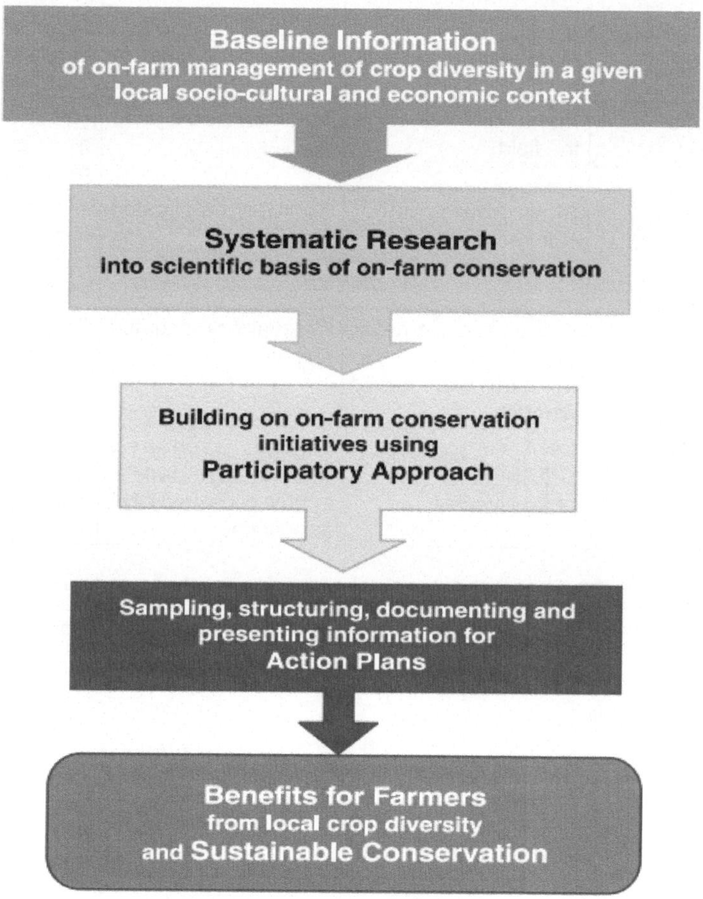

On-farm Conservation

Table 18.2: A Brief Description of the Various Approaches for Germplasm Conservation

Approach	*Accession Maintained as*	*Applications*	*Remarks*
In situ conservation	Growing plants in reserves protected for human interference	Germplasm of wild forms and wild relatives	Established in centres of origin/diversity; difficult to maintain
Ex situ conservation	Plants, seeds, shoot cultures *etc.*, in gene banks	Cultivated germplasm, wild forms and wild relatives	Established in places with appropriate facilities
(i) Seed banks	Seeds	All species producing orthodox* seeds	The most preferred approach
a. Base collection	Seeds with 5 per cent moisture at -20°C	The ultimate source of active collections	Disturbed only for-regeneration** and germination tests
b. Active collection	Seeds with 5 per cent moisture at close to0°C	Used for evaluation, multiplication and distribution	Periodically replaced by accessions from the base collection
c. Working collection	Seeds with 10 per cent moisture at -15°C	Used in breeding programmes	Maintained by the breeders
(ii) field banks	Plants maintained in the field	Asexually propagated species	Requires large area and considerable resources
(iii) slow growth cultures	Slow growing shoot cultures	Asexually propagated species	A substitute for field bank; requires tissue culture facility; in use for some species
(iv) Cryo-preservation	Cell, tissues, organs, *e.g.,* embryos, keep at -196°C in liquid nitrogen	Asexually propagated species	Requires considerable expertise and sophistication; in use for some species
(v) DNA banks	DNA fragments in COSMID clones or phage lysates	Extinct or nearly extinct species; a source of desired DNA sequences	Highly sophisticated; needs considerable resources

* Seeds that can be dried to 5 per cent or lower moisture level without any adverse effect on their viability.

**Growing the stored seeds in the field, and using the freshly harvested seeds for further storage.

Utilization

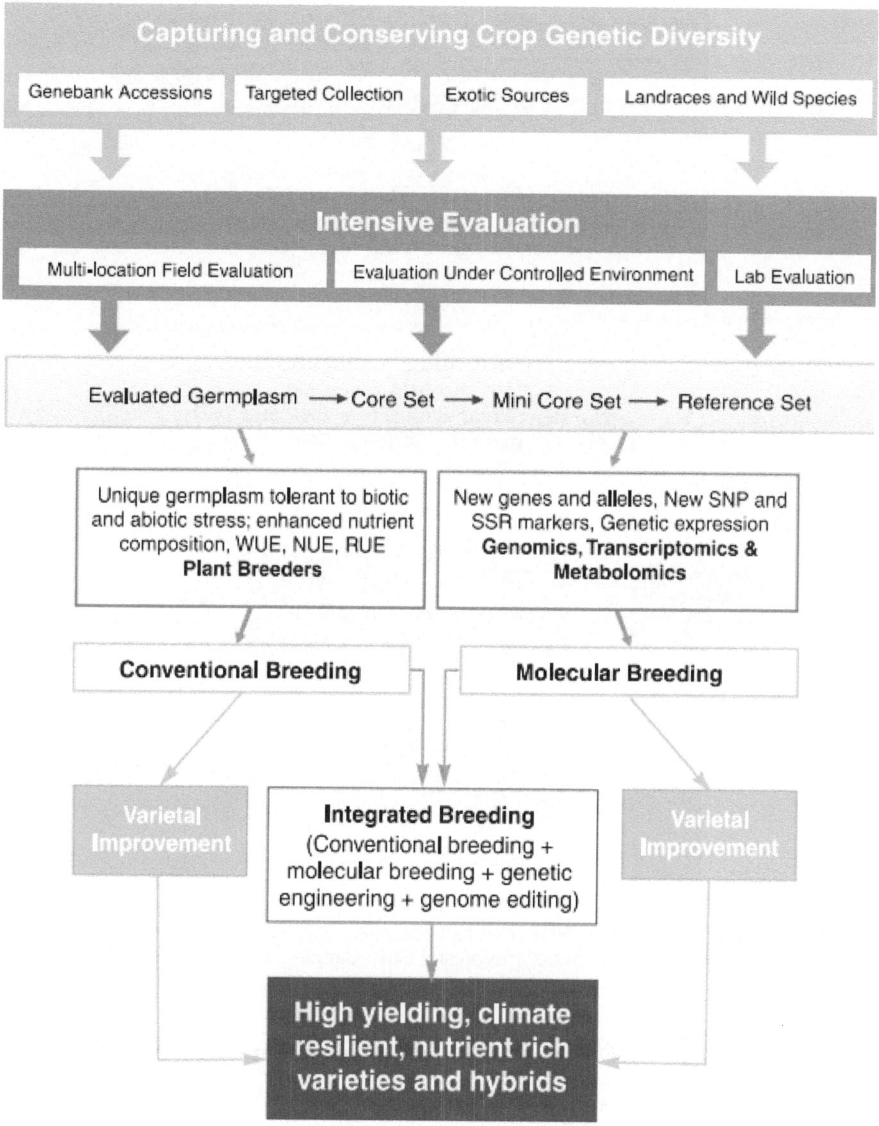

Enhanced Utilization of PGR for Sustainable Agriculture

WUE: Water use efficiency; **NUE:** Nutrient use efficiency; **RUE:** Radiation use efficiency; **SNP:** Single nucleotide polymorphism; **SSR:** Simple sequence repeat marker

Generation and Utilization of Genomic Resources

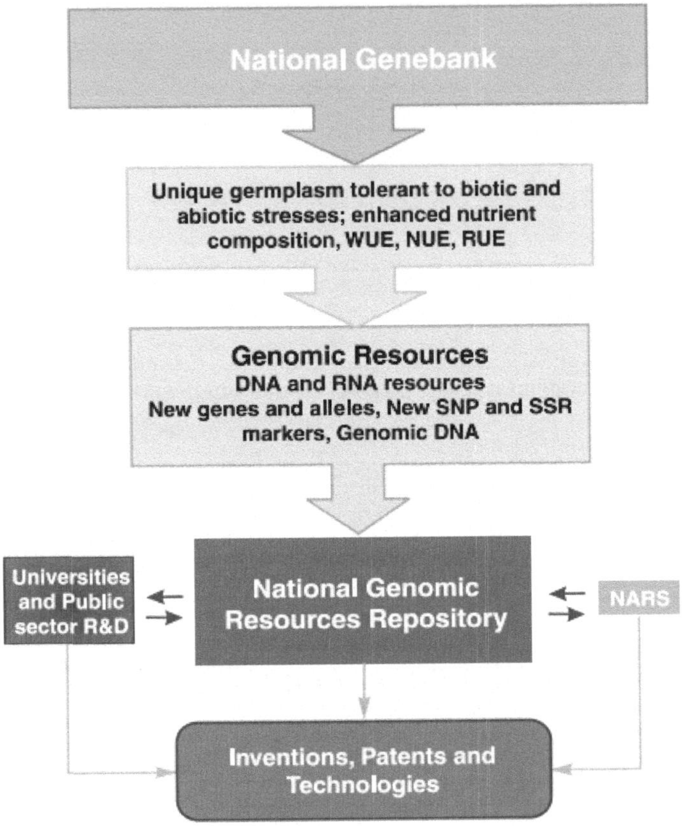

Knowledge and IP Generation

WUE: Water use efficiency; **NUE:** Nutrient use efficiency; **RUE:** Radiation use efficiency; **SNP:** Single nucleotide polymorphism; **SSR:** Simple sequence repeat marker

National Bureau of Plant Genetic Resources (NBPGR)

NBPGR headquarters New Delhi, is a nodal agency for activities and services related to plant genetic resources in India.

Brief History of NBPGR

The establishment of the Bureau coincided with the advent of the Green Revolution and was in response to the realization of perceived effects of the Green Revolution on agrobiodiversity. Further, it was in accordance with the international developments in the form of establishment of the International Board for Plant Genetic Resources (IBPGR), Rome, in 1974 (now renamed as International Plant Genetic Resources Institute).

1935 – Realization of need by the Crops and Soil Wing of the then Board of Agriculture and Animal Husbandry to establish an organization to carry out plant introduction activities.

1937 – Emphasizing the importance of crop germplasm by late Dr. B.P. Pal in his classic paper 'Search for New Genes'.

1941 – Reiteration of need of plant germplasm conservation by the Indian Society of Genetics and Plant Breeding.

Initiation by Dr. B.P. Pal to set-up a unit for assembly of global germplasm under phytosanitary conditions.

1946 – The ICAR Scheme for Plant Introduction commenced functioning in the then Botany Division of IARI under the leadership of late Dr. Harbhajan Singh.

1956- Expansion and strengthening of the 1946 scheme as Plant Introduction and Exploration Organization in the Botany Division at IARI, New Delhi.

1961 – Establishment of a separate Division of Plant Introduction, headed by Dr. Harbhajan Singh at IARI, New Delhi.

1970 – Constitution of a High-Level Committee by Govt. of India and input of visionaries like Dr. B.P. Pal, Dr. M.S. Swaminathan and Dr. A.B. Joshi to strengthen plant germplasm management activities.

1976 – Upgradation of Division of Plant Introduction as an independent institute *i.e.* National Bureau of Plant Introduction.

1977 – Renaming of National Bureau of Plant Introduction as National Bureau of Plant Genetic Resources (NBPGR), located at New Delhi.

Mandate

1. To plan, organize, conduct and coordinate plant explorations for collection of genetic diversity with particular reference to native and naturalized crops and their wild relatives in India and in specific cases abroad.
2. To undertake introduction, distribution, exchange and quarantine of plant germplasm for research purposes.
3. To characterize, evaluate and document germplasm collections of agri-horticultural crops and their wild relatives. Develop and operate National Database for storage and retrieval of information on plant genetic resources, farmer's knowledge, innovations, practices and ethno-botanical knowledge.
4. To promote safe conservation of plant genetic resources on a long-term basis using ex- situ methods and in-situ on farm conservation of landrace diversity (agro-biodiversity) in collaboration with other national/international organizations.

5. To conduct basic researches in the area of biosystematics, germplasm conservation and characterization for providing a sound scientific backup to Bureau's services and to develop the National Programme on Plant Genetic Resources.
6. To undertake teaching and training on plant genetic resources for postgraduate students linked with PG School, IARI; to organize suitable training programmes at the regional, national and international levels. To act as sole repository of all notified/released varieties along with herbarium specimens with relevant information and original/parental lines.
7. To conduct plant quarantine by examining seed and plant propagules under exchange for the presence of associated pests and pathogens and also to salvage healthy material from the infested/infected/contaminated samples
8. To develop and operate national data base on farmers knowledge, innovations, practices and ethnic knowledge related to agrobiodiversity and also for storage as well as retrieval of information on plant genetic resources

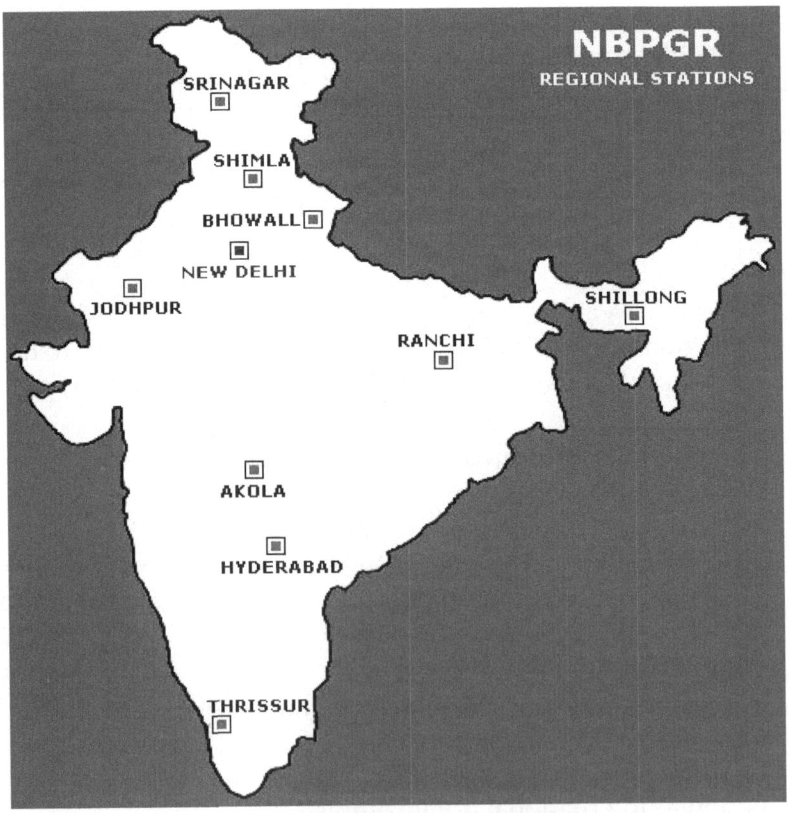

Regional Stations of NBPGR.

Table 18.3: According to the Last Updated data on: 28/05/2022 by NBPGR

NBPGR Regional Station	*Establishment Year*	*Area of Work*
Regional Station, Akola	1977	to meet the demands of Plant Genetic Resources activities in parts of Southern districts of Madhya Pradesh, hilly tract of Satpura, Maikala ranges, central-peninsular India, especially Maharashtra and northern Karnataka, Goa, Gawilgarh, plain cotton belt of Berar, Daman and Diu. works as satellite centre of Akola station.
Regional Station, Bhowali	1986	This was established initially as exploration base centre at Almora, working for germplasm collection in Kumaon and Garhwal hills. But in April, 1986 it was shifted to Bhowali and merged with Wheat Research Station. The mandate area for this is Central Himalayan Region.
Exploration Base Centre, Cuttack	1986	It was established in the campus of Central Rice Research Institute first. The potential area of collection is Odisha, West Bengal, parts of Chattisgarh, Jharkhand and humid tropical east coastal region. The area of collection has extended from Chhota Nagpur plateau to Koraput, southern most place of odisha covering Ganjam, Mayurbhanj, Keonjhar, Phulbani and Kalahandi districts.
Regional Station, Hyderabad	1977	Initially It was established in 16 acres land of Acharya N G Ranga Agricultural University Rajendranagar, Hyderabad as Plant Quarantine Station. It was initiated with an objective of Plant Quarantine clearance work in five mandate crops of ICRISAT and international trial material of paddy received from IRRI, Philippines.
Regional Station, Jodhpur	1976	I was established in the campus of Central Arid Zone Research Institute at Jodhpur, Rajasthan as a sub-station of Plant Introduction Division of Indian Agricultural Research Institute with an objective to collect the indigenous germplasm for arid/semi-arid conditions and to acclimatize the Sourced genetic material from abroad. The work area includes Gujarat, Rajasthan, and parts of Haryana.
Regional Station, Shillong	1978	This was established to cater the needs of germplasm conservation of all north-eastern states *i.e.*; Manipur, Meghalaya, Tripura, Mizoram, Assam, Sikkim, Nagaland, Arunachala Pradesh.
Exploration Base Centre, Ranchi	1988	It was established to cater the needs of systematic exploration of germplasm for evaluation and maintenance of germplasm of several tropical fruits and field crops native to states of Bihar, Jharkhand, West Bengal, Uttar Pradesh, Chhota Nagpur and potential adjoining region.

NBPGR Regional Station	Establishment Year	Area of Work
Regional Station, Shimla	1960	Initially it was established under Botany department, IARI as plant introduction station but further transferred under supervision of NBPGR in 1976. The station is primarily in charge of managing and conserving the plant genetic resources of the western Himalayas present in Himachal Pradesh and Jammu and Kashmir as well as in maintaining the gene bank for temperate fruits and recently introduced fruits. Being the largest germplasm collector and National Active Germplasm Site (NAGs) of crops like French bean, buckwheat and amaranth as well as temperate fruits, the station provides the facility of medium-term storage of orthodox seeds up to 12-15 years.
Regional Station, Srinagar	1988	This aims at germplasm collection of temperate fruits, rice, vegetables, millets, aromatic plants, medicinal plants from potential area of Jammu and Kashmir.
NBPGR Regional Station, Thrissur, Kerala	1977	It was established in the Kerala agriculture university near pineapple Research station for exploration and data collection of germplasm. The southern states of India are the designated region of this station which includes Tamil Nadu, Kerala, Karnataka, Goa, Pondicherry, Andaman and Nicobar Islands.

Table 18.4: Germplasm Collections are Maintained other than NBPGR

Sl.No.	Institute	Germplasm Crop
1.	IARI, New Delhi	Maize
2.	Central Potato Research Institute, Shimla (Himachal Pradesh)	Potato
3.	Central Institute for Cotton Research, Nagpur (Maharashtra)	Cotton
4.	Central Plantation Crops Research Institute, Kasargod	Plantation Crops
5.	Central Tobacco Research Institute, Rajahmundry (Andhra Pradesh)	Tobacco
6.	Central Tuber Crop Research Institute, Thiruvanthapuram (Kerala)	Tuber Crops other than Potato
7.	Central Rice Research Institute, Cuttack, Orissa	Rice
8.	Directorate of Oilseeds Research, Hyderabad	Oilseeds
9.	Directorate of Wheat Research, Karnal (Haryana)	Wheat
10.	Indian Grassland and Fodder Research Institute, Jhansi (Uttar Pradesh)	Forage and Fodder Crops
11.	Indian Institute of Horticultural Research, Bangalore	Horticultural Crops
12.	Indian Institute of Pulses, Kalyanpur, Kanpur (Uttar Pradesh)	Pulses
13.	National Research Centre of Groundnut, Junagarh (Madhya Pradesh)	Groundnut
14.	National Research Centre for Sorghum, Hyderabad	Sorghum
15.	National Research Centre for Soybean, Indore (Madhya Pradesh)	Soybean
16.	Sugarcane Breeding Institute, Coimbatore (Tamil Nadu)	Sugarcane

Convention on Biological Diversity (CBD)

CBDwas first agreement signed in 1993 by 173 countries including India. This agreement recognises the rights of nation over their genetic resources. Any country wishes to get access to PGR from other country must obtain PIC (prior informed consent) from authorities and MAT (Mutually agreed term) emphasises the condition under which PGR can be used.

International Treaty on Plant Genetic Resources for Food and Agriculture

The Headquarters is present at Rome, Italy.The Commission on Genetic Resources for Food and Agriculture was established in 1983, and the International Undertaking on Plant Genetic Resources was implemented on a voluntary basis. In 1996 at the Leipzig International Technical Conference on Plant Genetic Resources it took huge step for the adoption of the Global Plan of Action. All of this effort culminated in the historic adoption of the International Treaty on Plant Genetic Resources for Food and Agriculture in 2001, which came into effect on June 29, 2004. Mr. Kent Nnadozie is the present secretary of International Treaty on Plant Genetic Resources for Food and Agriculture.

International Treaty on Plant Genetic Resources aiming at establishing a global system to give farmers, plant breeders, and scientists access to plant genetic materials; assuring that recipients share benefits they derive from using these genetic materials with the countries they were originally from; and acknowledging the immense contribution that farmers make to the diversity of crops that feed the world. The Multilateral system was the unique approach to access and share germplasms created by the Treaty. It includes 64 most significant crops (collectively account for 80 per cent of the food which we obtain from the plants) to the global gene pool which is readily accessible to potential users of Treaty's ratifying nations for breeding, research and training.

FAO Roles in Germplasm Conservation

The world depends on genetic resources of agriculture, fishery, forestry and food production to sustain a healthy population of wild species as well as to increase the productivity and quality of domesticated plant and animal populations. Food security and nutrition are thus centred on the preservation and sustainable use of natural resources. Creating and maintaining the use of a wide range of diversity, both within and between species, implies creating options to address upcoming difficulties. FAO contributes its service to international policy through Commission on Genetic Resources for Food and Agriculture and the International Treaty on Plant Genetic Resources for Food and Agriculture. The Commission which is the only intergovernmental forum for addressing biodiversity, oversees and directs the formation of routine global assessments, negotiates global strategies, standards of conduct, as well as other instruments relevant to conservation and sustainable use of resources.

CGIAR (The Consortium of International Agricultural Research Centre)

Gene banks are important component of plant genetic resources conservation regulated by CGIARs. CGIAR gene banks maintains diverse germplasms from different locations by collecting from primary gene pools and exchanging with traditional communities as well as domesticating wild germplasms to fulfil Sustainable development goals (SDG). This germplasm includes wild relatives, improved varieties, diversified landraces and breeding related research materials of some major crops. In coordination with crop trust CGIAR gene banks perform functions like quality and performance management, online report, audit, review and validation (Lusty *et al.*, 2021). Newly updated Number of accessions of some important crops in CGIAR gene banks are listed below in Table 18.5 (Source: Crop trust. CGIAR Gene bank platforms).

Table 18.5

Crops	*Centre*	*Total No. of Accessions by 2020*
Rice	Africa Rice and IRRI	153,955
Wheat and Maize	International Maize and Wheat Improvement Centre (CIMMYT)	147,842
Potato, sweet potato and tubers	International potato centre	18,156
Sorghum and millets	International Crops Research Institute for the Semi-Arid Tropics (ICRISAT)	129,034
Cowpea, legumes, banana, cassava and yam	International Institute of Tropical Agriculture (IITA)	34,774
Grain legumes and forage crops	International Centre for Agricultural Research in Dry Areas (ICARDA) and International Livestock Research Institute (ILRI)	147,656
Tree crops	World Agroforestry (ICRAF)	14,919p

Agreements (Article 15) signed in 2006 between each CGIAR Centre and Governing Body of International Treaty on Plant Genetic Resources for Food and Agriculture (Plant Treaty) realise the importance of the collections in international policy.

Gene banks play crucial role in storing germplasms where local communities store the seeds for sharing and planting in near future. Establishment of new gene banks were supported and approved by FAO Commission on Genetic Resources for Food and Agriculture (CGRFA). CGFRA also provides the principles, crucial points and thresholds of gene bank management. To assess the degree of accuracy of passport data connected to genebank accessions, PDCI is used as a key performance indicator. In order to gauge changes in data quality over time, the PDCI range, which is between zero and ten, is periodically observed.

Conclusion

Plant genetic resources are the heritage of nature which are used for further development of varieties or hybrids based on breeding objectives. Plant genetic resources are the source of genetic variability which is fundamental requirement of any breeding programme. Uniformity in breeding lines has decreased the genetic basis of current cultivars due to which when cultivar is exposed to stresses (biotic or abiotic) results in huge yield reduction by hampering plant growth. On the other hand, sustainable utilisation of PGR should be considerate.

For proper collection, conservation and management of PGR, a suitable regulatory system is formed which undertake all the activities accompanied to PGR. Farmers contributed enormously for the ongoing development of plant genetic resources conservation. These regulatory systems help to acknowledge their effort, to protect their traditional knowledge and ensure their participation in decision making regarding the use of genetic resources. They also ensure the breeding of all crops (both major and minor) for maintaining diverse farming systems. The responsibilities of regulatory bodies can be best epitomized as; a crucial element in ensuring the preservation and sustainable use of genetic resources that enable the production of sufficient food to feed the world's expanding population.

REFERENCES

1. 3FAO's role in genetic resources. https: //www.fao.org/genetic-resources.com. 6th Feb, 2023.
2. https: //www.genebanks.org/resources/genebanks-innumbers/key-performance. 2nd Feb 2023.
3. NBPGR. https: //www.nbpgr.ernet.in. 2nd Feb, 2023
4. Lusty, C., van Beem, J., and Hay, F. R. (2021). A Performance Management System for Long-Term Germplasm Conservation in CGIAR Genebanks: Aiming for Quality, Efficiency and Improvement. *Plants*. *10*(12): 2627.
5. http: //www.nbpgr.ernet.in/About_NBPGR/Regional_Stations.aspx. 4th Feb,2023
6. Overview on International Treaty on Plant Genetic Resources for Food and Agriculture. https: //www.fao.org/plant-treaty/overview. 5th Feb, 2023
7. Roa, C., Hamilton, R. S., Wenzl, P., and Powell, W. (2016). Plant genetic resources: needs, rights, and opportunities. Trends in plant science, 21(8) : 633-636.
8. Singh. B. D, Plant Breeding principles and Methods. Tenth Reprinted Edition 2018, Kalyani Publishers, 36-39

Chapter 19

Biodiversity Conservation: Maintenance and their Use in Agriculture

Deepak Saran, Gyanisha Nayak*, Vivek Kumar Sandilya and Rohit Kumar

Department of Genetics and Plant Breeding, Indira Gandhi Krishi Vishwavidyalaya, Raipur – 492 012, Chhattisgarh
****e-mail: gyanishanayak@gmail.com***

ABSTRACT

The aim of this book chapter is to aware a growing researcher and farmer of the concept behind the conservation of the plant genetic resources, their maintenance, and use in the further improvement program. Despite these, diversity also improves soil health and the dynamic equilibrium of the agro-ecosystem. To review the current status of agricultural biodiversity, identify gaps in our information base, and propose future research and development actions, diversity knowledge is required. The common goal is to boom the sustainable use of biodiversity in improving humans' well-being and food and dietary security. The series scope is all components of agricultural biodiversity, starting from conservation biology of genetic assets to aware the farmers through agriculture extension sports, social sciences policy, and criminal components. Crop range presents the farmers with an extra choice within the production of a selection of vegetation in a given place to make bigger production-associated sports on numerous vegetation and also to convey down the feasible hazard in agriculture. It also covers the fields of research, information management extension education, communication and coordination, and knowledge sharing with the grower. It enables the farmers to grow surplus merchandise for sale at the marketplace and as a consequence enables to gain improved earnings to meet the opposite wishes associated with household nicely-being. Plant genetic resources are the important thing aspect of any agricultural production system and economic development and play a sizable position in global food safety. They underpin the capability of agriculture to cope with climate change and socio-economic conditions and enhance the use of bundles of variability "landraces cultivar" by their conservation and use. Biodiversity conservation is an important formula to conserve their natural agricultural biodiversity to achieve sustainable agriculture production improve the livelihood of growing farmers and global food security.

Keywords: *Biodiversity, Plant genetic resources, Germplasm conservation.*

Introduction

Plant Genetic Resources (PGR) are naturally available plant genetic material that has potential value to cope with this changing climate scenario and meet the food demand of the growing population. They consist of a large number of variables within the plants that leads to natural and human selection. They provide the raw material (genes) as germplasm or their wild relatives which are utilized to produce new and improved plant varieties and are an irreplaceable source of such characteristics *viz.* disease and pest resistance, adaptation, and high yielding with improved nutritional factors. Plant genetic resources are now and will continue to be used in the future, of inestimable value independently of whether breeders use them by employing conventional plant breeding and modern genetic engineering or non-conventional methods. The erosion of these plant genetic resources threatens to world food security. PGRs are a limited and perishable component of biodiversity. Some aspects in recent years have been causing rapid and extreme plant genetic erosion *viz.* uses of new technologies, the replacement of local or landraces varieties with improved ones, deforestation, and grazing, changes in crop cultivation methods, *etc.* This affects each cultivated species and masses of wild ones which have direct, oblique, or capacity for agricultural use. The erosion of these genetic assets should cause the extinction of precious material which has not but been exploited within the future. To maintain to boom the crop production and nutritional quality of food lies via the safety and green utilization of herbal plant genetic resources; this calls for their conservation, evaluation, documentation, and change. The International Treaty on Plant Genetic Resources for Food and Agriculture (ITPGRFA) becomes active on 29 June 2004 it becomes effective. This Treaty offers a framework for the international community to collaborate on issues like agrobiodiversity management, climate change adaptation, and food security while always keeping in mind the needs of farming communities, the hungry and underprivileged, and their right to food. For the sharing of all crop germplasm, the ITPGRFA supports a standard material transfer agreement. ITPGRFA is a thorough international agreement that supports Convention on Biodiversity (CBD). Through the protection, exchange, and sustainable use of

PGR for food and agriculture around the world, ITPGRFA seeks to ensure food security (PGRFA). The Global Action Plan (GPA) supports ITPGRFA and CBD.

As a defence against a well-known unpredictable future, there are commitments needed to preserve and use this content. PGFRA, or plant genetic resources for food and agriculture, is the source of the raw materials used by farmers and plant breeders to increase the quality and production of crops. The promotion of the utilisation of genetic resources for commercial purposes and as a component of biotechnology are some of the key reasons for conservation in order to ensure the future adaptation of elite lines and wild relatives. Additionally, it protects the existing natural diversity employed in crop breeding programmes as well as the information and qualities necessary to guarantee sustainable agriculture output.

More than ten thousand years ago, PGR was used for the first time. The genetic diversity seen in wild species and landraces was used by farmers to create superior cultivars with high yield. To maintain each species' wide-ranging genetic variety and the capacity to use it in a hybridization programme to produce variability rather than selection, it is necessary to protect wild germplasm and landraces.

Some Institutes at the Global and National Levels Related to Plant Genetic Resources

IBPGR (International Board for Plant Genetic Resources) managed the germplasm or plant genetic material at the international or global level, it was established in headquarters in **Rome, Itlay**. IBPGR is an international scientific organization under the aegis of the Consultative Group on International Agricultural Research (CGIAR) established in 1974. In 1991 IBPGR changed into IPGRI (International Plant Genetic Resources Institute). At the end of 2006, INIBAP (International network for the improvement of Banana and Plantain) and IPGRI jointly adopted the name Biodiversity International. At the national level NBPGR (National Bureau of Plant Genetic Resources) managed plant genetic resources. It was established in 1976 with its headquarters in **New Delhi.**

CBD known informally as the biodiversity convention is a multilateral treaty. It is an agreement between countries based on natural and biological resources. The convention was opened for signature at the Earth Summit in Rio de Janeiro on 5 June 1992 and enforced into 1993. The convention has three main goals including the conservation of biological diversity, sustainable use of its components, and equitable sharing of benefits arising from genetic resources

Some important points: Naturally existing biodiversity is 14 million spp. of plants expected by CBD and Estimated plant spp. around 0.33 million among that Described plant spp. Are 0.27 million. No. of mega-diverse countries are 17 among the world. Plant diversity found in India is 11.8 per cent among the 33 per cent endemic plant species and 49 endemic crops are found in

India. There are 326 wild relatives of cultivated crops. The Grassy stunt virus-resistant varieties of rice developed from *Oryza nivara.* The wild progenitor of the cultivated rice (Oryza *sativa*) is *Oryza nivara*is, it *is collected* from the Gonda district of Uttar Pradesh. There are 427 endangered plant species listed by BSI (Botanical Survey of India), Kolkata. 141 genera belonging to over 47 families are endemic in India (Higher plants). Highest endemism in India: Himalayan > Peninsular region>Andaman and Nicobar. India holds half of the world's aquatic floral diversity

Some Steps were taken by Govt. to Conserve Natural Genetic Resources

Biodiversity bill passed in Rajyasabha in 2002. Govt. Set up a board this is Biodiversity board placed in New Delhi and the Biological Diversity Act turned into surpassed in 2002. The biological diversity Authority was established in Chennai Northern hemisphere is rich in agro-biodiversity while the southern hemisphere is poor in the world. Agrobiodiversity hotspots in the country = 22. First gene bank in 1940's Saint Petersburg (now Leningrad) in Russia 40,000 seed samples at a stated time.

- NBPGR is a nodal agency for a single window exchange.
- The national gene bank was established in India at NBPGR in 1983.
- National gene bank was established in the USA at NBPGR in 1958.
- The national gene bank was established in Ghana at NBPGR in 1964.
- The national gene bank was established in Japan at NBPGR in 1966.
- National gene bank was established in Canada, Germany, and Italy at NBPGR in 1970.
- The national gene bank was established in Poland at NBPGR in 1971.
- The national gene bank was established in Turkey at NBPGR in 1972.
- Sec- 41 leads to the promotion of conservation sustainable use and documentation of biodiversity at the local level.
- PPVFRA stands for Protection of Plant Variety and Farmers Rights Act. PPVFR Act – 2001, turned into enacted to provide for the status quo of an powerful device for the protection of plant range, and the rights of farmers and plant breeders. It is encouraging the development and cultivation of latest sorts of flora. PPVFR rules were notified in 2003. PPVFR **Authority** was established in 2005. PPVFR regulation was notified in 2006. Headquarters of PPVFR Authority in New Delhi. PPVFR Act covered 18 species of 14 crops. The period for protection given for trees and vines crops under PPVFR Act 2001 is 21 years. The period for protection given for crops under PPVFR Act 2001 is 18 years.

- ☆ A variety that is released and granted a certificate of plant variety protection under the legal status of a given country is referred to as a protected variety. The right is granted by the govt. to a plant breeder to exclude others from producing or commercializing the propagating material of that variety for a minimum period of 15-20 years is known to plants breeder's right. The right is granted by the govt. to an inventor to exclude others from imitating, manufacturing, using, or selling the invention in question for commercial use during the specified period is called a patent.
- ☆ The researcher's right under PPVFRA is given in section 30.
- ☆ No. of Agro biodiversity hot spots identified by the PPVFR Authority was 22.
- ☆ The nodal agency for implementing the PPVFR Act is the Ministry of Agriculture, GOI.
- ☆ The chairman of PPVFR Authority is Dr. K.V. Prabhu in 2020-21.
- ☆ Plant Variety Journal of India is published bimonthly by PPVFR Authority.
- ☆ Protection of Plant Variety and Farmer's Rights Act protects farmers and plant variety.
- ☆ PPVFRA Bill (2001), Act is based on benefit sharing and compensation to farmers.
- ☆ CBD 29 December 1993 FIRST legal mechanism dealing with biodiversity.
- ☆ IUPGR stands for International Undertaking on Plant Genetic Resources, 1983.
- ☆ Material Transfer Agreement (MTA) issued by CBD in Bonn guidelines.
- ☆ To access the genetic resources CBD Article 15 is established.
- ☆ For incentive measures CBD Article 11
- ☆ To in-situ conservation CBD Article 8(j)
- ☆ CBD Article 15.7 to benefit sharing with parties
- ☆ CBD Article 10(c) sustainable use of a component of biological diversity.
- ☆ CBD Article 16 for Access to and transfer of technology.
- ☆ CBD Article 19 for the handling of biotechnology and distribution of its benefits.
- ☆ CBD Article 20 and 21 for financial resources and mechanism.
- ☆ CBD Article 22 for the relationship with other international connections.
- ☆ SMTA (standard material transfer agreement) released by ITPGRFA.
- ☆ SMTA is applicable for 64 crops in annexure-1.

- Any variety having (GURT- Genetic Use Restriction Technology) is barred from IPR protection in India.
- 4th International technical conference on PGR hosted by FAO at Leypzip, Germany, 1996.
- SAARC stands for South Asian Association for Regional Cooperation is established in Dec. 1985 with headquarters in Kathmandu (Nepal). It is an association of south countries in which Bangladesh, Bhutan, Nepal, Pakistan, Maldives, Srilanka, and Afghanistan added as a member of SAARC. The purpose of this organization is the development of member states in respect of collective, economic, technical, social, and cultural.

WTO (World Trade Organization) was set up on 1 January 1995, with headquarters- in Geneva, Switzerland. It is regulating the global exchange between nations. The outcome of the failure of negotiating governments ITO (International Trade Organization) become GATT. The first discussion of GATT changed into at some point of the United Nations Conference on Trade and Employment. After this, the signature of 123 countries in Marrakesh on 14, April 1994 of the Uruguay Round Agreements mounted the World Trade Organization. It is the successor to the GATT.

Geographical Indications of Goods (Registration and Protection), Act exceeded in 1999. GI Act is a sui generis Act of the Parliament of India for the safety of geographical indication in India.GI Tag guarantees that none aside from the ones registered as legal customers are allowed to use famous product names. Darjeeling tea have become the first GI Tagged product in India in 2004-2005, for the reason that then 323 items were added to this list. Initially, GI Tag has registered for 10 years length after this it can be renewed. Geographical Indications of Goods, Rules got here in 2003. Recently, in 2017 GI Tag became given by way of the Andhra Pradesh authorities to the Banganpalle mango variety. This range is known for its sweetness.

Components of PGR

Collections of genotypes or populations from cultivars, genetic stocks, wild species, *etc.* that are maintained as seeds, plants, tissue cultures, populations in the wild, or populations on farms are known as crop germplasm collections. Functionally, PGR includes crop plant relatives that are weedy or wild, improved/advanced cultivars, and landraces (either non-domesticated or domesticated). Although landraces still function as genetic reserves, they are quickly disappearing as more modern cultivars take their place. Advanced (current and obsolete) cultivars are resources that should be protected because they may be valuable to future breeders. Important sources of disease and pest resistance, as well as physiological adaptations not present in domesticated species relations, can be discovered in the wild and weedy relatives of crop species. The preservation of each of these is important to humanity because they are all gravely endangered with extinction. In general, changes made in response to farming have altered the genetic diversity in cultivated plants, which is derived from wild ancestor species. The changes also take into account the environment's physical, biological, cultural, and socioeconomic elements. The very unique nature of various settings has led to this domestication's many "ecospecific" adaptations, which have led to the emergence of landraces that are adapted to local environments.

Germplasm Activities

There are six important major actions related to plant germplasm:

1. Exploration and collection,
2. Conservation,
3. Evaluation,
4. Documentation,
5. Distribution, and
6. Utilization.

Exploration and Collection

Collection trips and collections that involve exploring the genetic diversity from numerous sources and gathering them all in one location make up exploration. Exploration is the process of seeking out new knowledge or resources. Exploration is a behaviour shared by all non-sessile animal species, including humans. It is an intriguing topic of study that has in the distant past drawn a sizable number of adventurers, naturalists, travellers, and plant hunters.

Sources of Collection

There are following five major sources of germplasm collections: *viz.*, 1) Gene banks, 2) Centre of diversity, 3) Gene sanctuaries, 4) Farmer fields and 5) Seed companies. Furthermore, collections can be done through local exploration trips to the crop diversity. Among that some are explained here.

Gene Banks

Genetic resources are preserved via gene banks. The most fundamental job of a gene bank is to maximise a fresh sample's vitality while maintaining its quality. The samples, also known as accessions, are monitored to make sure they aren't deteriorating. Regeneration is a crucial part of gene bank activities. Plant samples must be regularly grown out, given time to regenerate, and then harvested for new seed because they will eventually deteriorate even under the best conservation conditions. To preserve and revitalise genetic resources, gene banks must first collect them. Gene banks aren't just there to keep genetic resources safe; they're also there to make sure that they're utilised, whether that's through breeding programmes, research facilities, or farmer's fields. To accomplish this, it is crucial to check that the collections are appropriately defined and recorded, and that the needed material is available. The information systems used by gene banks are becoming more and more important to researchers and breeders seeking information on the distribution of crops and their wild relatives. The Tropical Botanical Garden and Research Institute (TBGRI), Palode, and TVM for medicinal plants and spices. The National Bureau of Plant Genetics Resources, New Delhi, for crop plants and medicinal plants. For medical plants and spices, there is the Central Institute for Medical and Aromatic Plants (CIMAP) in Lucknow, as well as an animal research facility called the Central Drug Research Institute.

Wildlife Sanctuaries: As long as they don't have a detrimental effect on the wellbeing of the animals, human activities including wood harvesting, obtaining minor forest products, and private property rights are allowed in sanctuaries. A sanctuary is a place that is strictly protected and dedicated to the preservation of animals. The boundaries of sanctuaries are not clearly defined, and tourism and other forms of controlled biological intrusion are permitted.

Merits and Demerits of Exploration and Collection of Germplasm

Merits

1. Collection aids in exploring crop genetic diversity and bring together in one place.
2. It decreases the loss of genetic material due to genetic erosion.
3. Sometimes, we get desirable genetic sources during germplasm exploration trips.

Table 19.1: Some Important Germplasm World Collection

Germplasm World Collection	*Crop*
Senegal	Groundnut
Beltsville, USA	Small grain crops
Cambridge, U.K.	Potato
Florida	Sugarcane
Ethiopia, Africa	Coffee
New Zealand	Sweet potato

Demerits

1. Collection of germplasm especially from other countries, sometimes there is a danger to the entry of new diseases, new insects, and new weeds.
2. In remote areas, the collector sometimes has an encounter with animals like elephants, rhinos, tigers, lions, and snakes which involve life.

The collection is a tedious job. The collection of germplasm has to be done generally from uncultivated areas like hills, mountains, river valleys, and forests, where the collector faces many problems *viz.* boarding, lodging, and transportation. Transportation of huge collections also faces difficulties in the exploration and collection.

The major activities of germplasm enhancement may be summarized as follows:

- Preventions of genetic uniformity lead to reduce genetic vulnerability.
- Increasing yield in major world food crops *viz.* rice, wheat, and sorghum, were accomplished through introgression of unadapted desirable genes (*e.g.*, dwarf genes).
- To introduce new and desirable quality traits (*e.g.*, starch, protein).
- Introgression of disease-and insect-resistance genes.
- To introduce new environment-resistance genes (*e.g.*, drought resistance).

Centre of Origin

Areas that played a role in the emergence or diversification of a certain species or population are commonly referred to as "Centers of Origin." The starting point is the location where a group of organisms, both domesticated and wild, first developed their distinctive characteristics. Vavilov initially identified a region of the world as a Vavilov Centre (1926), also known as the Vavilov Centre of Diversity, as an actual hub for the domestication of plants. A theory was suggested by Vavilov at the locations where cultivated plants were first created. He claimed that domestication of plant life did not occur randomly around the globe but rather in specific locations.

- Primary center of origin: crop plants originated from more diverse wild species found in the region.
- Secondary centre of origin: while not being the primary source, some regions exhibit a significant diversity of forms.

Eight main and 3 sub-centre of origin were originally proposed by Vavilov in 1926:

1. China
2. Hindustan
   - Indo –Burma
   - Malaya java
3. Asia Minor
4. Central Asia
5. Abyssinia

6. Mediterranean region
7. South America
   - ☆ Peru
   - ☆ Chile
   - ☆ Brazil -Paraguay
8. Central America

Note: USA centre of origin also consider as a centre of origin, newly.

Germplasm Conservation

In situ and ex-situ preservation of germplasm are the two main methods.

In situ Conservation

In this strategy, plant genetic resources are safeguarded in their natural habitat or on the origin site. It refers to the preservation and restoration of populations of species in their native habitats and, in the case of domesticated or cultivated species, in the contexts where their distinctive qualities have originated. It also clearly refers to the preservation of natural habitats and ecosystems. It can also be described as conservation taking place naturally or ordinarily within a species' natural range. For instance, natural parks, sanctuaries, and reservoirs.

Ex-situ Conservation

In Ex-situ conservation, germplasm doesn't preserve within the natural locations of starting place however below the supervision of breeders at expert websites called germplasm or gene banks. Ex-situ conservation method, "Off-site conservation". It is the technique of keeping an endangered plant or animal via disposing of a part of the population from a habitat that is in hazard and moving it to an exclusive area, which may be in the wild or below human supervision. Ex-situ conservation includes both more modern, really contentious laboratory strategies in addition to a number of the oldest and most famous conservation strategies. The established system of gene banks, which includes genetic sources facilities, zoos, botanical gardens, lifestyle collections, and many others, is used to carry out ex-situ conservation, which uses sample populations.

Ex-situ conservation serves many purposes, including the rescue of threatened genetic material, creating research materials for conservation biology, increase the amount of germplasm for storage in various types of ex situ facilities, provide materials for diverse uses to ease or lessen the pressure of wild amassing, grow the species with recalcitrant seeds that can't be kept in a seed store; material made to order for conservation education and display; create materials for management, habitat restoration, reinforcement, and reintroduction.

Depending on the lifetime of seeds or on their viability during storage, seeds can be categorised into two groups;

1. **Orthodox Seeds:** These seeds can be dried to a 5 per cent or underneath five per cent moisture content without lack of their seed viability. These are frost-resistant and lengthy-time period saved seeds. Mostly orthodox seeds come from annual temperate crop species *viz.* Wheat, rice, barley, pearl millet, different cereals pulses, and oil seed vegetation. They contain moisture content material of 30–50 per cent at physiological maturity at the time of maturity.
2. **Recalcitrant Seeds:** Recalcitrant seeds are short-term stored seeds that cannot be safely dried to a moisture level below 30 per cent. These can't survive colder temperatures and are challenging to maintain because of their high moisture content, which encourages microbial infection and causes seeds to lose viability quickly. They are made up of fruits and the seeds from plantation crops including cocoa, coffee, and coconut.

Types of Germplasm Collections

Plant genetic resources are maintained by four types:

1. **Base collections:** These collections are kept in long-term storage arrangements rather than being assigned to researchers. The term "predominant collections" or "a wholeseries of genetically diverse species" is used to describe them. It keeps the germplasm in its original state. Depending on the species, germplasm is stored at lower humidity + 5 per cent, temperatures (10–18°C), or cryogenic (150–196°C). Materials can be kept for many decades in the proper conditions.
2. **Backup collections**: It is responsible to supplement the base selection. It is a duplicate collection to ensure provides germplasm whenever disaster comes.
3. **Active collections:** Plant breeders are given access to active collections for research purposes. These are preserved at temperature below 15°C (often near 0°C), have an 5 per cent moisture content, and are used for 10 to 15 years. These collections are utilised for dissemination, reproduction, and evaluation.
4. **Breeders or working collections**: The working collection is of utmost importance for plant breeders in the crop improvement program. Breeder's or working collections are known for short-term storage for 3-5 years and it stored at less than 15°C and preserve at low moisture content +10 per cent

A core collection consists of a set of the minimum number of accession that together represents the genetic diversity of the concerned crop and its wild relatives. Therefore, each accession in a core collection is to some extent, representative of several accessions in the gene bank. However, a core collection should not replace the original collection.

Gene Pool Concept

Primary Gene Pool (GP1)

- ☆ It includes strains of the species in which hybridization between the GP1 is easy without any difficulty and resultant F_1 hybrids are generally more vigorous than parents. They show normal meiotic chromosome pairing, normal gene segregation, and complete seed fertility.

Secondary Gene Pool (GP2)

- ☆ Cross between GP1 and GP2 exhibit some degree of difficulty, resulting in F1 Hybrid exhibiting partial fertility and partial Cross difficulty due to differences within crossing individuals at chromosome ploidy level, chromosome alterations, or genetic barriers

Tertiary Gene Pool (GP3)

- ☆ Germplasm of GP3 exhibit a high range of difficulty to cross with GP1 and F1 hybrid exhibits complete sterility.

Quarter Gene Pool (GP4)

- ☆ Member of GP4 includes '**Transgene**'. It is also called the **"Ocean of germplasm"**. In this Cross between the individuals does not possible.

Table 19.2: Some Important Active Germplasm Collections Maintained in India at Institutions other than NBPGR, New Delhi

Sl.No.	*Institutions*
1	Central Institute for Cotton Research, Nagpur
2	Central Plantation Crops Research Institute, Kasargod
3	Central Potato Research Institue, Shimla
4	Central Tobacco Research Institute, Rajahmundry
5	Central Tuber Crop Research Institue, Thiruvanantpuram
6	NRRI, Cuttack
7	Directorate of Oilseeds Research, Hyderabad
8	Directorate of Wheat Research, Karnal
9	Project Directorate on Maize, New Delhi
10	Indian Grassland and Fodder Research Institute, Jhansi
11	Indian Institute of Horticultural Research, Banglore
12	Indian Institute of Pulses Research, Kanpur

Sl.No.	*Institutions*
13	Sugarcane Breeding Institute, Coimbatore
14	Indian Institute of Vegetable Research, Varanasi
15	National Research Centre for Groundnut, Junagarh
16	National Research Centre for Soybean, Indore
17	National Research Centre for Sorghum, Hyderabad

Protection of Plant Variety and Farmers Rights Act (PPVFRA), 2001

The Special Features of this Act are

It is an easy system to defend farmers' range. This act covers new plant types, essentially derived sorts, existing varieties, and farmers' varieties. Farmers' types and existing types don't need innovation, but they do need protection from the "terminator gene" and the genetic use restriction age. recognising farmers as growers' equals in terms of innovation, conservation, breeding, and plant variety preservation. It permits unique arrangements for profit-sharing and the establishment of gene funds for the preservation of agricultural biodiversity. When a material is supplied, if it no longer performs as expected overall, particular provisions are made to make up for the farmer's loss. Farmers get their expenditures covered. Failure to provide quality material to farmers at a reasonable cost is established a basis for mandatory licencing and the formation of a tribunal.

Farmers' Rights

Except in cases where the sale is made for reproduction under commercial cultivation, farmers have the right to conserve, use, exchange, share, or sell their farm products of protected varieties. The rights that result from protecting, enhancing, and making genetic resources accessible are also included in the farmer's right. Additionally, it refers to the legal claims that result from farmers' past, current, and future efforts to preserve, enhance, and make available plant or animal genetic resources, particularly those in areas of origin or diversity.

Researchers' Rights

As per this right "Protected varieties will be freely available for undertaking experiments or legitimate research projects. It is not prohibited for anyone to use a variety as the starting point for developing further variations. However, if a protected variety is used repeatedly as parental material in a commercial production or marketing arrangement, breeder permission will be required".

Benefit Sharing

It refers to the portion of the benefit that a claimant will be entitled to, as assessed by the Authority, that is gained by a breeder of a cultivar from an agent or a licensee of the cultivar, as the case may be.

National Gene Fund

In order to conserve genetic resources, a National Gene Fund will be established and funded through earnings from benefit-sharing agreements, royalties, community compensation, and contributions. Farmers that participated in the process of genetic resource protection were rewarded by this money. The Fund will be put to use through appropriate plans to increase their share to entities or people, compensate village communities, maintain and conserve plant genetic resources, and develop varieties.

Compulsory Licensing

Any person may apply to the Authority at any time following the passing of three years following the registration of a variety, claiming that the public's reasonable needs for seeds or other propagating material of the variety have either not been met or are either not available at a reasonable price, and requesting the granting of a compulsory license to produce, distribute, and sell the seed or other propagating material of the variety.

- ✰ Royal botanical lawn situated at Kew, England, organized the advent of quinine and rubber trees from south America into India.
- ✰ Beltsville, USA collects the germplasm of small grain crops at the world level.
- ✰ Bambey, Senegal collects the germplasm of groundnut
- ✰ Canal point, Florida sugarcane
- ✰ FRI stands for Forest Research Institute, Dehradun in 1906.
- ✰ NBPGR has five substations for testing the introduced plant material
    1. Shimla, Himachal Pradesh
    2. Jodhpur, Rajasthan
    3. Kanya kumari, Tamilnadu
    4. Akola, Maharashtra
    5. Shillong, Meghalaya

Following are some recommendations we would like to offer in light of the need to diversify agriculture in light of climate change and the rapid deterioration of natural resources, as well as the current state of diversification in our nation.

- ✰ The government should act right away to promote crop diversification in the nation by increasing funding for research and development, offering effective support services and credit institutions, marketing, processing, and packing agricultural products.

☆ NGOs and Panchayats Raj organisations might be active in educating farmers about the commercialization of agriculture and the existence of various government-sponsored programmes for the purpose and advantages of crop diversification.

Table 19.3: Eight Main and 3 Sub-center of Origin

Centres	*Crops*
China	Soybean, Peach, citrus sinensis
Hindustan	Rice, brinjal, mung bean
Central Asia	Bread wheat, lentil, carrot, radish, chickpea
Asia Minor or near eastern centre	Einkorn wheat, durum wheat, melons, grapes, almond
Mediterranean	Durum wheat, lettuce, broad bean, cabbage.
Abyssinian	Barley, lentil, pea
Central America	Corn, common bean, upland cotton
South American	Sweet potato, potato, tomato, papaya, tobacco
Subsidiary centres	
Indo-Malayan	Banana, coconut, yam
Chile	Potato
Brazilian –Paraguayan	Peanut, rubber tree, pineapple, cocoa.

Conclusions

The fundamental objective of biodiversity conservation is to conserve the natural genetic resources, and maintenance of broad-based genetic diversity to improve potential value to ensure their availability for exploitation by present and future generations. It maintains the existence of extinct species which aids in existing natural variability and creates variability through a hybridization program. It helps the breeder to make the selection for crop improvement programs and develop high-yielding cultivars with good nutritional quality. It supports meeting the requirement for world food security and the economic development of growing farmers. There is a need to extend the basic knowledge of conservation, maintenance, and their use of wild spp. and landraces through agricultural extension activities to the growing farmers and researchers for the further improvement program.

REFERENCES

Cruz-Cruz, C.A., González-Arnao, M.T. and Engelmann, F. (2013). Biotechnology and conservation of plant biodiversity. Resources, 2(2), pp.73-95.

De M. 2017. Field gene banks: the living repositories of plant genetic resources, Harvest (Online), **2**: 21-27.

Dhillon B. S and Saxena S. (2003). Conservation and Access to Plant Genetic Resources.

Government of India. (1994). Conservation of Biological Diversity in India: An Approach. MoEF, GOI, New Delhi pp 48.

Kumar A., Sharma A., Balyan U.K. and Dhaka T.S. (2015). Economic Botany and Plant Genetic Resources- At a Glance, pp 32-70.

Singh, B,D. 2015. Domestication and germplasm conservation, Plant Breeding Principles and Procedures, Kalyani Publishers, pp14-41.

Singh P. (2017). Fundamental of plant breeding, pp 18-61.

Index

A

B

C

D

E

F

G

H

I

J

L

M

T

W

Z

9 789359 191638

Printed at Repro India Ltd.